Boundary Management

Mitsuru Kodama

Boundary Management

Developing Business Architectures for Innovation

 Springer

Mitsuru Kodama, Ph.D.
Professor of Information and Management
College of Commerce and
Graduate School of Business Administration
Nihon University
5-2-1 Kinuta Setagaya-ku
Tokyo 157-8570
Japan
kodama.mitsuru@nihon-u.ac.jp

ISBN 978-3-642-03788-7   e-ISBN 978-3-642-03789-4
DOI 10.1007/978-3-642-03789-4
Springer Heidelberg Dordrecht London New York

Library of Congress Control Number: 2009939280

Cover design: WMX Design GmbH, Heidelberg

Printed on acid-free paper

Springer is part of Springer Science+Business Media (www.springer.com)

To my family

Preface and Acknowledgement

In recent years, the development of ICT and digitization has increased the need to develop new products and services and build business models that transcend industries and merge different technologies. Technology innovation in the past closely pursued and developed specialist knowledge, but with the development of unprecedented new products and services based on new concepts, innovations increasingly arise from merging one technology field with another. Amid continuous environmental change, dynamic strategic management to deliberately and constantly create new positioning (including new products, services, and business models) and values is an important theme for practitioners on a day-to-day basis.

How should companies exploit and implement strategy under a dynamically fluctuating environment? What is the essence of dynamic strategic management? These issues are common points of deliberation for strategy researchers and numerous corporate leaders alike. The research question I would like to pose as a specialist in the fields of innovation and strategic management is that of how to achieve this corporate strategy for dynamic strategic management. This book suggests a framework and case studies for dynamic strategic management theory for strengthening existing business and taking new positions to target new business (products, services, and business models) under a rapidly changing environment.

The essence of strategic management goes beyond companies simply adapting to environmental change while creating appropriate strategies for the future. It also involves companies optimizing the individual management elements that comprise the corporate system (including organization, strategy, operation, and leadership) in alignment with these factors, and achieving continuance and growth through integrative and dynamic development. How companies consider congruence with the environment and dynamically transform corporate boundaries to adapt to the environment (or create new environments) has become a key theme in the implementation of corporate strategy. In this book the optimal design of a corporate system comprising the management elements of strategy, organization, operation, and leadership aimed at designing corporate boundaries compatible with the environment is referred to as "business architecture."

"Business architecture" is a concept for optimizing corporate boundaries aimed at realizing targeted business models and corporate system design involving stakeholders. To optimize the corporate boundaries, companies must partially and wholly optimize the individual management elements (strategy, organization, technology, operation, and leadership) comprising the corporate system that has achieved congruence with the environment. The management to optimize these corporate boundaries and the corporate systems comprising individual management elements is referred to in this book as "boundary management."

The concept of "boundaries congruence" inside and outside the corporate system, and the building of optimal architecture concerned with environmental change and with management elements such as strategy, organization, technology, operation, and leadership are key to realizing dynamic strategic management. This book presents the concept of "business architecture" and optimizing processes as a corporate system based on multiple corporate case studies.

I could not have completed this book without interacting painstakingly and rigorously with practitioners. I would like to extend my gratitude to these practitioners, who are too numerous to count. I would like to express my appreciation, especially, for the senior executives and managers of the companies targeted (including Sony, NTT-DATA, NTT-DoCoMo, Toyota, Honda, Omron, Takara, Recruit, First Retailing, Panasonic, and Canon) for this book's research. I would also like to express my deep appreciation to Dr. Ikujiro Nonaka, emeritus professor at Hitotsubashi University, for a wealth of intellectual stimulation including innovative thinking and philosophical insights.

Concerning the publication of this book, the author wishes to extend his appreciation to Dr. Niels Peter Thomas, Senior Editor at Springer, who provided tremendous support. Finally, I would like to express my gratitude for a grant-in-aid for scientific research from JSPS, for the publication of this book, and for publishing grant-in-aid from Nihon University's College of Commerce.

Nihon University, Tokyo   Mitsuru Kodama

Contents

1 Dynamic View of Strategic Management 1
1.1 Dynamic Competitive Environments and New Business Models . 1
    1.1.1 New Competition and Strategic Positioning 1
    1.1.2 Business Ecosystems as New Competitive Rules 3
1.2 The Need for Dynamic Strategic Management Theory 5
1.3 Strategic Management as Practice Process 5
1.4 Boundaries Congruence Through Boundary Management 7
1.5 The Aims and Structure of This Book 10
References 13

**2 Theoretical Framework of Dynamic Strategic Management
Through Boundary Management** 15
2.1 The Optimal Design of Corporate Boundaries 15
2.2 The Concept of Business Architecture 16
    2.2.1 Strategy as Business Architecture 18
    2.2.2 Technology as Business Architecture 18
    2.2.3 Organization as Business Architecture 19
    2.2.4 Operation as Business Architecture 22
    2.2.5 Leadership as Business Architecture 22
    2.2.6 Dynamic Process of Business Architecture 22
2.3 Strategic Management in a Dynamic Environment 24
    2.3.1 Positioning- and Resource-Based Views 24
    2.3.2 Environment Adaptive Strategy 25
    2.3.3 Environment Creation Strategy 30
2.4 Managing Corporate Innovation Streams 31
References 33

**3 Developing New Business Models Through Dynamic
Boundary Management: Case Studies of Sony and NTT-DATA** . . . 37
3.1 Innovation by Internal Corporate Venture (ICV) 37
3.2 Sony: A Case Study 38
    3.2.1 Starting Up a New Business 38
    3.2.2 Integrating Knowledge from Different Fields
           Through Network Concepts 42

      3.2.3   SWS and SWS Network Creation   43
  3.3   NTT-DATA: A Case Study   46
      3.3.1   New Business Model Development   46
      3.3.2   Building Flattened Organizations and Organic Networks .   47
      3.3.3   Construction Innovations   50
      3.3.4   Marketing Innovations   52
      3.3.5   Content Innovations   54
  3.4   Gathering and Integrating Distinct Creative and Practical Knowledge   56
  3.5   Optimizing Boundaries Congruence for Business Architecture . .   58
      3.5.1   Optimizing Boundaries Congruence in SCE   58
      3.5.2   Optimizing Strategic Framework Through
             Boundaries Congruence in NTT-DATA   60
  References   61

**4   Developing New Broadband Services by Dynamic
   Collaboration Through Strategic Boundary Networks: A
   Case Study of NTT DoCoMo**   63
  4.1   Innovation Through Synthesis of Exploration and Exploitation . .   63
  4.2   Case Study: Mobile Phone Business Innovation   65
      4.2.1   DoCoMo's Innovations   66
      4.2.2   Phase 1 (1992–1998): The Challenge of Voice
             Communication   66
      4.2.3   Phase 2 (1999–2004): The Challenge of Multimedia   69
      4.2.4   Phase 3 (2004 to Present): The Challenge of a
             Lifestyle Tool   78
  4.3   Forming Small-World Networks Within the Company   81
  4.4   Network Integrative Competences Through Leadership Teams . .   83
  4.5   The Dynamic Boundaries Congruence of Business Architecture . .   85
  References   90

**5   New Knowledge Creation Through Leadership-Based
   Strategic Community**   93
  5.1   Introduction   93
  5.2   Networks of Strategic Communities   93
  5.3   Summary of an In-depth Case   96
      5.3.1   Current Status and Issues in the Field of Veterinary
             Medicine in Japan   96
      5.3.2   Formation of Strategic Community and Networked
             Strategic Communities   98
      5.3.3   Innovation in New Knowledge Creation 105
  5.4   Results and Discussion 106
      5.4.1   Characteristics of Networked SCs 106
      5.4.2   Synthesizing Capability Through Dialectical
             Leadership of Community Leaders 108
  5.5   Managerial Implications: Toward the Realization
      of Strategic Community-Based Organizations 110

    5.6   Conclusion   112
    References   112

6 New Theoretical Framework and Insights Derived from
    Comparative Case Studies   115
    6.1   New Business Creation by Transformation   115
            6.1.1   The Importance of Change Management   115
            6.1.2   Issues in Promoting New Business: Non-
                       congruence Among Individual Management
                       Elements   116
    6.2   Comparative Case Analysis and New Insights   121
            6.2.1   Congruence of Strategy and Technology   121
            6.2.2   Congruence of Strategy and Organization   122
            6.2.3   Congruence of Strategy and Operation   123
            6.2.4   Management and Leadership   123
    6.3   Crucial Elements for Boundaries Congruence
            and Successful Innovation   125
            6.3.1   Creative/Productive Dialogue and Practitioners'
                       Recognition Capability at SWS and Networked SWS . . .   125
            6.3.2   Drawing in Knowledge Boundaries and the
                       Practice Process at Process-Based Organizations   128
            6.3.3   Traffic and Synthesis Among Dual Networks   131
    References   139

7 Theoretical and Managerial Implications   141
    7.1   Boundaries Management Frameworks   141
    7.2   Optimizing Organization Architecture for Innovation   145
    7.3   Business Architecture for Integrated Organization   147
    7.4   New Knowledge Creation From the Process-Based Organization .   150
    7.5   Network Architecture Thinking   151
    7.6   Network Architecture of the Small-World Structure   153
            7.6.1   Internal Network Architecture of SWS   154
            7.6.2   The External Network Architecture of SWS   155
    References   159

8 Conclusion   161
    8.1   The Dynamics of Business Architecture   161
    8.2   Innovation from Network Architecture Thinking   162

Appendix: Research Methodology and Data Collection   165

References   167

Index   169

Chapter 1
Dynamic View of Strategic Management

1.1 Dynamic Competitive Environments and New Business Models

1.1.1 New Competition and Strategic Positioning

The business environment surrounding companies in the twenty-first century is changing ever more dramatically. Managers face multiple issues, including business globalization, technological innovation, the maturing of markets, price competition, and environmental problems. In industries with mature markets, the skill of business domain strategy divides winners from losers, and companies that win through enhance their relative corporate value, even in times of low growth. It has been noted that the companies achieving sustainable growth in the world markets in recent years have the following points in common. First, they possess core business through sound core competence (e.g., Hamel and Prahalad 1989), and effectively exploit and sustainably promote these core competences (e.g., Zook 2003). Second, they advance into peripheral businesses (e.g., Zook and Allen 2001) and different sectors that enable them to demonstrate synergies with core businesses. Third, they build new value chains focusing on the customer (e.g., Kim and Mauborgne 2005).

There are many examples of success for the first two points. Seiko Epson has steadily expanded peripheral domains exploiting related technology from its core competence of watches and clocks, and has grown from a small regional company to a company outstanding in its industry. In the post-war period, Sharp developed radio and TV technologies on the platform of the core consumer electronics market. On this foundation the company has come to redefine and grow stronger businesses by exploiting its superiority in each key domain. Today Sharp is a world leader in the development and sale of liquid-crystal TVs and solar batteries. Canon, meanwhile, planned business expansion on the basis of its core technologies of lens development and production, moving into the product areas of photocopiers, printers, digital cameras, and medical equipment.

Major Japanese goods distribution company Yamato Transport entered the home delivery business for small-lot items, an area where profits were difficult to reap and industry competition was intense. With a painstaking focus on service, Yamato

M. Kodama, *Boundary Management*, DOI 10.1007/978-3-642-03789-4_1,
© Springer-Verlag Berlin Heidelberg 2010

Transport succeeded in differentiating itself from other companies. Seven Eleven also succeeded in creating business in a new field when it focused on customer convenience with the opening of ATMs (financial business) in geographically separate stores, thus expanding its business domain from the distribution industry to finance. Meanwhile, the US company Dell established firm leadership of its core business of direct PC sales. For a time Dell advanced into store retail selling, withdrawing after failures, but later it expanded its consumer market centered on corporate business and grew to be a global business.

IBM also planned to switch to a solutions and service business based on conventional computer development and sales. D&B was already the overwhelming leader in its core credit information business, but judged that much potential remained, and methodically reinvested to double sales and triple operating profit ratio over the course of a decade. The Service Master started out as a carpet cleaner before succeeding in its long-term strategy of capturing peripheral domains closely related to its core business. After 15 years sales have increased five-fold, and the company is maintaining an ROE of more than 20 percent.

The games business of the Japanese company Nintendo is a prominent example illustrating the third point above. The case of Nintendo's new DS and Wii games machines is a classic example of how to create a customer model from a "no-customer" model. Nintendo pioneered the new "Blue Ocean" (e.g., Kim and Mauborgne 2005) market by providing games for children and adults who had previously played very little.

In recent years, however, new competition has arisen from the increasing ambiguity of core business boundaries, the dramatic expansion of peripheral business domains (Zook and Allen 2001), and the merging of businesses in different fields (Kodama 2007a). As a result, thinking about how to form new markets and redefine one's own core business competences is becoming increasingly important. Key expansion strategies might include merging businesses adjacent to core businesses, building areas that are separate yet connected to the core businesses, or merging one's own and another company's core businesses to cultivate new business in a different area. The new focus of strengthening such core businesses by allowing core competences to accumulate while cultivating new business domains is growing important to corporate leaders and managers. Cultivating new peripheral domains and business in other fields leads to the discovery and cultivation of new, highly independent strategic positions, and is also an outstanding strategy for constantly pursuing and maintaining new positions (Markides 1999).

Concepts like Blue Ocean-type strategies (e.g., Kim and Mauborgne 2005) that cultivate new markets by presenting new customer-focused value propositions to major customer segments and strategies that capture uncultivated customer segments with products using low-end disruptive technologies (Christensen 1997; Christensen and Raynor 2003) also become important ideas for acquiring market position through building new value networks. To cultivate these kinds of peripheral domains and businesses in other fields, corporate leaders and managers must possess the "peripheral vision" to perceive diverse signals (Day and Schoemaker

2005), and prepare to enter industries outside those of the company's core business domains. Examples include the case of Apple, whose music distribution business is threatening the core business of the established CD industry.

In recent years the Internet, mobile phones, and various digital products (such as the i-pod and game machines) exploiting ICT (Information and Communication Technology) have appeared. Used as product and strategy tools in every industry and business situation, they are spreading globally and continuing to create new business models in diverse fields. With fast-changing industries such as ICT and digital products, practitioners must base their strategic thinking on different competitive rules.

1.1.2 Business Ecosystems as New Competitive Rules

Thus the competitive environment accompanying the move to ICT and digitalization is spawning new competitive rules. In the second half of the twentieth century, Chakravarthy (1997) defined the four clusters of the turbulent Infocom industry (communication providers, communications support, information processors, and information providers), and indicated the importance of leader companies focusing on diversifying and strengthening core competences while acquiring a competitive advantage in the highly complex and rapidly changing infocom industry. He noted that innovation leader companies must become "repeat first movers" in order to manage the network effect (e.g., Shapiro and Varian 1998) through positive feedback and build new value chains. This does not mean, however, that all the ICT leader companies need to do is ride the wave of environmental change (markets and technologies) faster than the other companies. As mentioned later, the leader companies must promote not just competition but coordination and collaboration among stakeholders, and build a business ecosystem as an entire industry.

The rapid advance of ICT, for example, is transcending the merging of different technologies and industries, including the e-money business, IC tag solutions, the merged communications and broadcast business, and the shift of the music distribution business toward the i-Pod and mobile phones to form new competitive rules, business models, and business networks (e.g., Kodama 2007a, 2007b). The value chains of broadband and wireless business series comprising mobile phones, PCs, and other terminals; communications network platforms for billing, settlement, and e-money; applications such as distribution networks and IC tags; and content such as broadcasting, images, music, and advertising are being built through competition and collaboration among stakeholders.[1] These distinctive value chains have created business ecosystems comprising content business through mobile phones

---

[1] In the ICT industry, information distribution business models such as i-mode are built through co-evolution arising from coordination and collaboration among content and platform providers. For details see Kodama (2009).

(e.g., Moore 1993; Gawer and Cusumano 2004; Iansiti and Levien 2004). These markets are still expanding today while creating co-evolution models for mobile phone businesses.

According to Moore (1993), a business ecosystem is an economic community supported by a foundation of interacting organizations and individuals—the organisms of the business world. Moore suggested that a company should not be viewed as a member of a single industry but "as part of a business ecosystem that crosses a variety of industries," and highlighted the importance of co-evolving the capabilities amongst the business ecosystem members. With this kind of business ecosystem, companies and other organizations correspond to economic communities supported by interaction among diverse individuals and organizations. These economic communities produce value-laden goods and services for the customers (including a company's own vendors, partners, corporate customers, and competitors), and the customers themselves also become community members. The members within this community (groups of companies) co-evolve their own abilities and roles, and develop activities following the direction indicated by one or more core companies. The companies that show leadership may change over time, but the role of corporate leader of a business ecosystem continues to be evaluated highly by the community.

In Japan, the leaders of the mobile communications carriers centered on NTT DoCoMo, KDDI, and Softbank are pursuing new business models arising from co-evolution through coordination and collaboration among stakeholders, and new dynamic business ecosystems are being created as a result (Kodama 2009). This kind of new business network simultaneously creates new competition and cooperative relationships that differ from the conventional, established business structures.

The relationship between coordination and collaboration among business networks can also be seen not only among different industries but also in the same industry. The consumer electronics business, with its rapidly changing markets and technologies, is one such example. Examples include collaboration between Samsung Electronics and Sony for the development and production of liquid-crystal panels and joint ventures and comprehensive tie-ups (associations to create the strongest scenarios) between Matsushita Electrical Industrial Company (now Panasonic Corporation), Hitachi, and Canon for liquid-crystal displays to develop and produce and strengthen and develop cutting-edge technologies. In the background of this collaboration among competitors is each company's strategic purpose in sharing dynamic expertise and avoiding the risks that have been associated with the technology development costs of recent years.

Meanwhile, collaboration through strategic tie-ups with this kind of external partner company also becomes the trigger that enhances creativity (Kodama 2009). Companies can absorb external partners' ideas and expertise (Chesbrough 2003) and enhance the originality of their own products and services using connect and development strategies like the US company P&G. (Huston and Sakkab 2006) They do this by such means as scrutinizing not just their own but also other companies' knowledge and skills with open innovation and open business models (Chesbrough 2006). Absorbing customer competence also becomes important for strengthening the competitiveness of goods and services (Prahalad and Ramaswamy 2004;

Kodama 2002). When markets and technologies change dramatically in this way, coordination and collaboration among companies, including customers, become important elements in enhancing companies' mutual creativity and building new business models.

1.2 The Need for Dynamic Strategic Management Theory

Thus recent developments in ICT and digitalization have increased the need to develop products and services and build business models that merge different technologies and transcend industries. Technology innovation up to now has closely pursued and developed specialized knowledge, but the number of cases where some fields of technology merge with others to realize new products and services developed from unprecedented new concepts is increasing. Because of this, the dynamic view of strategy management whereby companies intentionally reposition themselves again and again with new products, services, and business models amid a constantly changing environment to form and create new value is a key theme in practitioners' daily lives.

What kind of strategies should companies exploit and execute amid this kind of dynamically changing environment? What is the essence of dynamic strategic management? These questions are faced by strategic management researchers and corporate leaders alike. The essence of strategic management goes beyond adapting to a changing environment and creating strategies appropriate for the future. It optimizes individual business elements such as organizations, technology, operation, leadership, and human resources aligned with these adaptations, and achieves corporate continuity and growth through comprehensive and dynamic development of such elements.

An important focal point is that strategy should constantly change and dynamically build and rebuild. Strategy is a serial and continuous activity for creating new values, and is also a dynamic process leading to a sustainable competitive edge for the company (Montgomery 2008). So what is strategic management (dynamic strategy management) as a dynamic process? The intention of this book is to inquire into the ideal shape of corporate strategy for this dynamic strategic management. Put another way, it is the issue of how to manage a company's various strategy streams.

1.3 Strategic Management as Practice Process

Academic strategic management research undertaken up to now has handled static research from detailed computational analysis with economic methodologies as an orthodox branch of management studies. Although various subtle analytical tools (such as SWOT analysis, PPM, the 7S model, and five force) have been developed to implement corporate strategy planning on a practical level, the applicable range of these tools with regards to competitive strategy is limited, and they have been unable to transcend the range of past and current strategic analyses. The exemplary Porter theory (Porter 1980; 1985) based on the economics of industrial organizations posits maintaining the optimal strategic position aimed at competition to distribute value as the essence of strategy. Even many researchers believe that strategies up to this point have been essentially plans, and that because they closely analyze environmental change and corporate resources, and then build corporate strategic positions on this basis, they provide the starting point for corporate strategies.

Meanwhile, focusing on practitioners' actual business activities revealed large gaps at the micro-level in the strategic planning and implementation stages. In many cases, events did not go as planned when using strategy analysis tools to closely analyze the environment, create plans, and implement them. One topic of strategic analysis is the points that were consistently unsuccessful regardless of the care taken over the plan. The essence of strategy research cannot lie solely in analysis prior to strategy implementation, nor in analysis of strategy-driven results or causes that are already past.

Kaplan and Norton (2008) indicate that strategy and operational breakdowns are the cause of stagnating corporate results. While it is vital to have in place a management system as a strategy manual 'of explicit knowledge to reinforce workplace activities and forge strong links between specific strategy aims and strategic tools supporting practical corporate activities, it is not the essence of strategy. The key focus is not only strategy planning but a comprehensive view including the "human system" comprising operations, technologies, and the actions and leadership of practitioners (including customers who are closely connected with the organization) working inside and outside the corporation. If the "human system" is ignored, the expectation of an outstanding management system will be pie in the sky. Strategy is an organically living object and really looks like "human process."

Jack Welch, one of the world's leading business practitioners, wrote in his book (Welch and Welch 2005):

I know that strategy is a living, breathing, totally dynamic game... In real life, strategy is actually very straightforward. You pick a general direction and implement like hell. If you want to win, when it comes to strategy, ponder less and do more. . . Obviously, everyone cares about strategy. You have to. But most managers I know see strategy as I do—an approximate course of action that you frequently revisit and redefine, according to shifting market conditions. It is an iterative process and not nearly as theoretical or life-and-death as some would have you believe.

In sum, strategy is a dynamic practice process that involves practitioners recognizing and managing the dynamic relationships among people through daily practice, producing concepts and ideas to build new values while developing corporate visions and strategy aims through trial and error, and then reliably implementing them. Then the new value creation activities arising out of this practice process deliver superior products and services while returning general satisfaction to the community. Amid the time-axis flow running from past to future, the essence of strategic management lies in dynamic practice that continuously pursues new strategic positioning with the aims of adapting to changes in the external environment (or creating environmental change oneself), realizing the ideal shape of the company, and creating new value. The focal points of "strategy as practice" and "strategic management as practice" (Johnson et al. 2007; Kodama 2007b) are also approaches that create new foci and insights incorporating practitioners' subjective aspects (examples might include practitioners asking what kind of strategies they should implement on a daily basis through the dynamic relationships of strategic activities among people) at a micro level. These aspects derive from objective macro-level frameworks comprising conventional analytical strategy viewpoints (e.g., Porter 1980) and process-driven strategy viewpoints (e.g., Mintzberg 1978).

1.4 Boundaries Congruence Through Boundary Management

Research regarding the correlation between strategy selection, planning, and results have produced a great deal of expertise, generally resulting from quantitative research arising from statistical methodology. The question of how to implement strategy under a dynamic environment is undeniably limited by the static, quantitative studies. From this standpoint, in order to understand more deeply the pros and cons of strategy and the causal relationship between the timing and process of strategy implementation and the results, it is necessary to advance further with penetrative qualitative studies regarding strategies and organizations that consider temporal fluctuations, and with field research emphasizing practitioner philosophies. I believe that a focus on the dynamic process view of strategic management, considering strategy from the viewpoints of future purpose, philosophy, and practice, will grow increasingly important. This approach would ask what kind of thinking is needed for top, middle, and lower management to position themselves with regard to strategy, organizations, and self-and-other amid a time scale stretching from the past to the future; what shape should the corporation take in the future; what kind of person I would like to be; what is the best situation for the people who work there; and what kind of existence is ideal? Put another way, the essence of strategic management is "the pursuit of future creation."

As a point of focus that captures the dynamic process of strategic management, researchers and practitioners must consider very carefully the dynamic changes of a corporate system (including, for example, internal corporate management elements such as strategy, organization, technology, operations, and leadership) that adapts to changes in the dynamic environment surrounding the company (markets, technologies, competition, co-operation, structures and other factors). The boundaries between the corporations and the environment and the corporate system (the corporate boundaries) define the relationship between the environment and the corporate business model. Environmental change gives rise to change of corporate boundaries, and its influence extends to the individual management elements within the corporate system. Conversely, the active or passive changes of the individual management elements within the corporate system give rise to changes in the corporate boundaries, and further extend their influence on the environment.

Put another way, for a company to adapt to environmental change (formulating and implementing environment adaptive strategy) or actively influence the environment to create a new environment (formulation and implementation of environment creation strategy) practitioners must deliberately modify the strategies, organizations, technologies, operations, leadership and the other management elements within a company, and plan congruence of boundaries among these management elements. A large number of practitioners, for example, face practical questions such as "What kind of organizations and operations can realize the targeted strategies?" "What elemental technologies are required to realize the targeted product and business models?" "What shape of corporate leadership and management is required to realize new strategies?" and "What kind of consciousness innovations do employees require?"

However, strategies, organizations, technologies, and other management elements incorporate heterogeneous contexts, while practitioners possess "thought worlds" (Dougherty 1992) and heterogeneous "mental models" based on different contexts and experiential knowledge. The focus on recognizing individual elements within the environmental and corporate system also differs from practitioner to practitioner. Accordingly, constraints among practitioners inevitably give rise to "knowledge boundaries" (Kodama 2007a) as barriers to congruence among these elements. Nevertheless practitioners driving innovation and reform management should perceive these not as constraining knowledge boundaries, but rather as triggers to create new knowledge and competences (in my experience of business, I found that capable practitioners tend to perceive these knowledge boundaries as opportunities).

A key focus of dynamic strategic management relating to the process of change in such environments and corporate systems is how practitioners perceive and acknowledge changes for each boundary (whether environment and corporate system boundaries or boundaries between individual management elements within the corporate system) and attempt to create boundaries congruence. In this book, I will consider the congruence of boundaries between environments and corporate systems and within corporate systems, and look at the connections among these elements. The foci will be on how corporate management should implement the process of changes in strategic management, aimed at practitioners realizing environment adaptive strategies and environment creation strategies in the context of fast-changing environments.

In order for companies to continually develop and grow, it is important for practitioners to create such processes (incremental and radical) of strategic management change on a temporal axis and build new business models. The concept of the above-mentioned boundaries congruence and its implementation is key to achieving this (see Fig. 1.1).

In this book I will refer to this management of boundaries congruence as "boundary management." In order to grasp the mechanism of boundary management, it is necessary to analyze and consider the key mechanism of the practitioners' "practice processes." These processes are not just the activities at practitioners' formal organizations within and outside the organization, but also the bridge process of

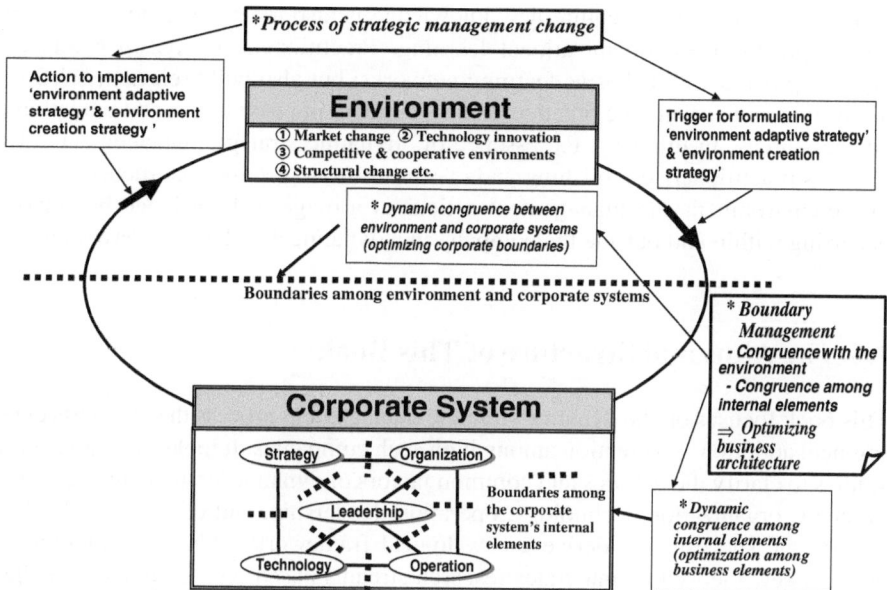

Fig. 1.1 Dynamic view of the strategic management process

knowledge boundaries arising from the formation of human networks (real and virtual space) that transcend organizational and corporate boundaries. The formation of human networks signifies network formations such as "Ba" (Nonaka and Konno 1998), the community of practice (Wenger 1998), and the "small-world structure (SWS)" (e.g., Watts and Strogatz 1998). "Boundaries congruence" is achieved by practitioners determining and implementing strategies that result from forming networks through practice process (known as "strategy practice").

In "Ba" and the community of practice," information, context, and knowledge is shared closely among members, and promotes daily best practice and organizational learning. In the community of practice, study that is generally within the same general organization or function and rigorous organizational study among the companies that make up the *keiretsu* network of the car industry (e.g., Dyer and Nobeoka 2000; Dyer and Hatch 2004) have given rise to the formation of highly interconnected cohesive network structures (Coleman 1988). In the mobile phone industry, meanwhile, companies intending to innovate through acquiring new knowledge may maintain weak relationships by building sparse networks with potential partners, and monitor new information as required (Kodama 2007a: Padula 2008). This kind of high-tech company implements bridging ties among different networks through shortcuts to potential partners (or partner groups forming clusters) aimed at acquiring new opportunities. The action of practitioners taking these shortcuts builds small-world structure (SWS) or small-world networks within or among organizations, and incorporates new innovation-oriented knowledge within the SWS or within each organization through the medium of the SWS.

Building an SWS to acquire new knowledge is also a new behaviour that transcends practitioners' organizational learning. Practitioners not only execute the
practice process with cohesive learning networks; but also build relevant, SWS for
innovation and execute the practice process challenging new topics and unexplored
business areas. In this way, the essence of "dynamic strategic management" also
involves practitioners establishing and executing strategies for congruence relating
to the environment and management actions at a range of knowledge boundaries
occurring within and outside the company in a dynamically changing environment.

1.5 The Aims and Structure of This Book

This book focuses on the dynamic strategic management process that supports environment adoption and creation among high-tech companies. It looks at detailed case
studies to clarify the features and common factors of dynamic strategic management
(strategy, organization, technology, operation, leadership, and other areas) leading
to strategic success, and derive a new logical framework. This book attempts to
create a new theory that integrates research streams based on the three established
pieces of the corporate "congruence model" (e.g., Tushman and O'Reilly 1997),
the "boundaries-based view of the firm" (e.g., Carlile 2004; Santos and Eisenhardt
2005), and "strategy as practice" (e.g., Whittington 2004) (see Fig. 1.2). The book
has three main features.

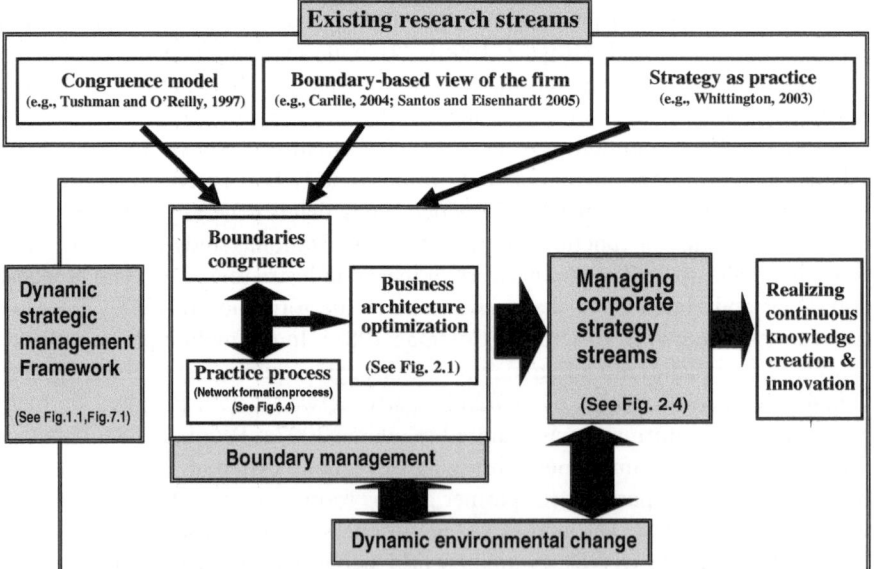

Fig. 1.2 Analytical framework for this book

The first is the derivation of a basic framework for dynamic strategic management from an integrated viewpoint including strategy, organization, technology, operation, and leadership under a dynamic environment. This approach differs from the traditional theoretical approach characterized by static analysis of markets and strategies. The second identifies "boundaries congruence" as the essence of dynamic strategic management, attempts strategy practice analysis from the new viewpoint of strategy and boundaries, and goes on to present the concept of business architecture to realize boundaries congruence. The third emphasizes dynamic process analysis while providing new knowledge relating to the dynamics of boundary management from the viewpoint of practitioners' practice process further incorporating black boxes, rather than the conventional in-house macro strategy-making process that progresses through established research (e.g., Minzberg 1978; Burgelman and Välikangas 2004).

This book also pays attention to the practice process comprising the thoughts and actions of practitioners inside and outside the organization as new insights that link these three features. The book suggests that the SWS (e.g., Watts and Strogatz 1998) formed by the practitioners (who form one of the key elements of this practice process) both inside and outside companies and their organically networked structures (networked SWS) become strategizing and organizing enablers supporting environment adoption and creation.

Put another way, the SWS organic networks integrate practitioners' constraints at the boundaries inside and outside the organization as well as at the knowledge boundaries, and initiate changes that cement the bridges between the boundaries. The "boundaries bridge" then achieves boundaries' congruence to create optimal business architecture, and practitioners implement environment adaptive and environment creation strategies (referred to as "corporate strategy streams" in this book) to promote a new knowledge creation process (see Fig. 1.2). This book also points to new implications and a future research agenda for this expertise.

Engaging with these research topics is both academically and practically significant. Since they have been somewhat overlooked in the past, I have attempted to address them directly with detailed field surveys in the business workplace. I have also attempted to pursue new management theories relating to dynamic strategic management. This book focuses on field research (interviews, ethnography, and participant observation) among Japanese companies (see Appendix: Research Methodology and Data Collection). It analyzes the features and competitiveness common to each company's dynamic strategic management process, and plans to create new corporate management concepts.

The structure of this book is explained below. Chapter 2 ("A theoretical framework of dynamic strategic management through boundary management") presents the concept of "corporate strategy streams," which are one of the basic frameworks of the dynamic strategic management process, and demonstrates that the dynamic synthesis of the environment adaptive strategy and the environment creation strategy is one of the creative sources of a company's sustained competitiveness. Then it considers boundaries congruencies among the environmental and corporate system and within the corporate system, which is important to initiate strategy streams, and presents a new concept of business architecture and boundary management framework.

Chapter 3 ("Developing new business models through dynamic boundary management: a case study of Sony and NTT-DATA") looks at three of Japan's leading cutting-edge companies and analyzes the factors behind the success of these companies' internal corporate ventures (ICVs) from the viewpoint of business architecture and boundary management.

Chapter 4 ("Developing new services by dynamic collaboration through strategic boundaries networks: A case study of NTT DoCoMo") takes up in-depth case studies of leading Japanese mobile telecommunications carrier NTT DoCoMo, which pioneered and commercialized the mobile Internet service i-mode to the world. It demonstrates how the strategy practice of dual network formation from the dynamic process of strategic management stretching from the company's foundation to the present has built optimal business architecture, and forms an important element in realizing corporate innovation streams.

Chapter 5 ("New Knowledge Creation through Leadership-based Strategic Community") provides new practical viewpoints in knowledge management and leadership theory of project management through an in-depth case study. It is argued that community leaders, particularly business community leaders, must recognize that a strategic community (SC) as "Small-World Structure (SWS)" comprises of diverse types of business and processes needed to achieve continuous business innovation. The community leaders serve an important function in creating a networked strategic communities (networked SWS).

Chapter 6 ("New theoretical frameworks and insights derived from comparative case studies") looks at several leading Japanese companies that have succeeded in achieving new lines of business (products, services, and business models), and analyzes these common elements of new business success from the viewpoints of boundaries congruence and boundary management.

Chapter 7 ("Theoretical and managerial implications") presents new knowledge acquired through the numerous in-depth case studies of previous chapters. The strategic management process which practitioners operating at a micro-level dynamically achieve environment adaptive and environment creation strategies comprises the management of "boundaries change" inside and outside the company or organization. Chapter 7 considers integrated organization and internal corporate venture (ICV) models to optimize environmental and internal corporate system congruences. It then demonstrates how the practitioners' network architecture concepts promote boundaries congruence and become knowledge creation enablers.

Finally, Chap. 8 considers the book's conclusions and touches on future research topics.

References

Burgelman, R. A., Välikangas, L. (2004). Managing internal corporate venturing cycles. *Sloan Management Review*, 46(4), 26–34.

Carlile, P. (2004). Transferring, translating, and transforming: an integrative framework for managing knowledge across boundaries. *Organization Science*, 15(5), 555–568.

Chakravarthy, B. (1997). A new strategy framework for coping with turbulence. *Sloan Management Review*, 38(2), 69–82.

Chesbrough, H. (2003). *Open Innovation*. Boston, MA: Harvard Business School Press.

Chesbrough, H. (2006). *Open business models: how to thrive in the new innovation landscape*. Boston, MA: Harvard Business School Press.

Christensen, C. M. (1997). *The Innovator's Dilemma: When New Technologies Cause Great Firms to Fail*. Boston, MA: Harvard Business School Press.

Christensen, C. M., Raynor, M. (2003). *The Innovator's Solution*. Boston, MA: Harvard Business School Press.

Coleman, J. (1988). Social capital in the creation of human capital. *American Journal of Sociology*, 94, 95–120.

Day, G., Schoemaker, P. J. (2005). Scanning the periphery. *Harvard Business Review*, 83(11), 135–148, November.

Dougherty, D. (1992). Interpretive barriers to successful product innovation in large firms. *Organization Science*, 3(2), 179–202.

Dyer, J., Hatch, , N. (2004). Using supplier networks to learn faster. *Sloan Management Review*, 45(3), 57–63.

Dyer, J., Nobeoka, K. (2000). Creating and managing a high-performance knowledge-sharing network: the Toyota case. *Strategic Management Journal*, 24(3), 345–367.

Gawer, A., Cusmano, M. A. (2004). *Platform Leadership*. Boston, MA: Harvard Business School Publishing.

Hamel, G., Prahalad, C. K. (1989). Strategic intent. *Harvard Business Review*, 67(3), 63–76.

Huston, L., Sakkab, N. (2006). Connect and develop inside Procter & Gamble's new model for Innovation. *Harvard Business Review*, 84(3), 58–66.

Iansiti, M., Levien, R. (2004). *The Keystone Advantage: What the New Dynamics of Business Ecosystems Mean for Strategy, Innovation, and Sustainability*. Boston, MA: Harvard Business School Press.

Johnson, G., Langley, A., Whittington, R. (2007). Strategy As Practice: Research Directions and Resources. Cambridge, UK: Cambridge University Press.

Kaplan, R., Norton, D. (2008). Mastering the management system. *Harvard Business Review*, 86(1), 62–77.

Kim, W. C., Mauborgne, R. (2005). *Blue Ocean Strategy*. Boston, MA: Harvard Business School Publishing.

Kodama, M. (2002). Strategic partnership with innovative customers: a Japanese case study. *Information Systems Management*, 19(2), 31–52.

Kodama, M. (2007a). *The Strategic Community-Based Firm*. Basingstoke: Palgrave Macmillan.

Kodama, M. (2007b). Knowledge Innovation –Strategic Management As Practice. Cheltenham: Edward Elgar Publishing.

Kodama, M. (2009). *Innovation Networks In Knowledge-Based Firm – Developing ICT-Based Integrative Competences*. Cheltenham: Edward Elgar Publishing.

Markides, C. (1999). *All the Right Moves: A Guide to Crafting Breakthrough Strategy*. Boston, MA: Harvard Business School Publishing.

Mintzberg, H. (1978). Patterns in strategy formation. *Management Science*, 24(4), 934–948.

Montgomery, C. (2008). Putting leadership back into strategy. *Harvard Business Review*, 86(1), 54–60.

Moore, J. (1993). Predators and prey: a new ecology of competition. *Harvard Business Review*, 71(3), 75–86.

Nonaka, I., Konno, N. (1998). The concept of "Ba": building a foundation for knowledge creation. *California Management Review*, 40, 40–54.

Paduda, G. (2008). Enhancing the innovation performance of firms by balancing cohesiveness and bridging ties. *Long Range Planning*, 41(4), 395–419.

Porter, M. (1980). *Competitive Strategy: Techniques for Analyzing Industries and Competitors*. New York: Free Press.

Porter, M. (1985). *Competitive Advantage*. New York: Free Press.

Prahalad, C. K., Ramaswamy, V. (2004). *The Future of Competition: Co-Creating Unique Value With Customers*. Boston, MA: Harvard Business School Press.

Santos, M., Eisenhardt, K. (2005). Organizational boundaries and theories of organization. *Organization Science*, 16(5), 491–508.

Shapiro, C., Varian, H. R. (1998). *Information Rules*. Boston, MA: Harvard Business School Press.

Tushman, M. L., O'Reilly, C. A. (1997). *Winning Through Innovation*. Cambridge, MA: Harvard Business School Press.

Watts, J., Strogatz, S. (1998). Collective dynamics of "small-world" networks'. *Nature*, 393(4), 440–442.

Welch, J., Welch, S. (2005). *Winning*. New York: Harper Business.

Wenger, E. C. (1998). *Community of Practice: Learning, Meaning and Identity*. Cambridge: Cambridge University Press.

Whittington, R. (2004). Strategy after modernism: recovering practice. *European Management Review*, 1(1), 62–68.

Zook, K. (2003). Beyond the Core: Expand Your Market Without Abandoning Your Roots. Boston, MA: Harvard Business School Press.

Zook, C., Allen, C. (2001). Profit from the Core: Growth Strategy in an Era of Turbulence. Boston, MA: Harvard Business School Press.

Chapter 2
Theoretical Framework of Dynamic Strategic Management Through Boundary Management

2.1 The Optimal Design of Corporate Boundaries

A company must actively change its own corporate governance structure and corporate boundaries and strengthen its strategic position under constantly changing environments (or new, self-created environments). Research relating to previous corporate boundaries indicates that the decision-making of the corporate governance structure and corporate boundaries relies on various elements including the transaction cost economic theory view, the capability and competence views, and identity.[1] Decision-making over the questions of what business activities to implement within the company, or whether to access external resources through market agreements with regard to building strategically targeted value chains, then become important elements of corporate strategy for large corporations and venture companies alike (e.g., Pisano 1990).

Santos and Eisenhardt (2005) suggest four specific elements (efficiency, power, competence, and identity) that determine corporate boundaries. These four elements of cost (efficiency), autonomy (power), growth (competence), and coherence (identity) in corporate activity are required as basic management factors, and are also important themes determining corporate boundaries. In recent years, the determining of corporate boundaries through strategic outsourcing aimed at cost reductions has further enhanced the efficiency of corporate activities. Moreover, the building of keiretsu networks rooted in long-term relationships of trust with subcontracting companies in Japan's automobile, consumer electronics, and telecommunications device industries is advancing influence through the power of corporate activity and the autonomy of the subcontracting companies.

The business models (game and solutions businesses, broadband services) arising from dynamic, cross-company organizational networks among Sony and NTT-Data described in Chap. 3, and the mobile phone business emerging from exploratory and exploitative activities through the formation of DoCoMo's dual organizational

---

[1] See Klein (1988) and Williamson (1975) for the transaction costs economics view; Penrose (1959), Nelson, and Winter (1982) for the capabilities and competences view; and Kogut and Zander (1992) for the identification view.

M. Kodama, *Boundary Management*, DOI 10.1007/978-3-642-03789-4_2,
© Springer-Verlag Berlin Heidelberg 2010

networks described in Chap. 4 create co-evolutionary business ecosystems among stakeholders, and greatly influence the boundaries of numerous companies and industries. Moreover, the identities of Japanese companies integrating management styles such as "value and resonance," "corporate culture," and "teamwork" distinctive among the leading Japanese companies Toyota and Honda have become resources creating both incremental innovation ("kaizen") and radical innovation (Shibata and Kodama 2009).

Considering congruence with the environment and the question of changing the corporate boundaries dynamically to adapt to an environment (or create a new environment) become important themes in corporate strategy implementation. Companies have to determine strategy aims for sustainably competitive products, services, and business models, and realize them by implementing optimal designs for both vertical (value chains to achieve company-determined strategies) and horizontal (expansion and diversification of business domains) corporate boundaries. The optimal design of corporate boundaries adapted to the environment requires optimal design of a corporate system structured from the management elements of strategies, organizations, technologies, operations, and leadership (described later). This optimal design of a corporate system congruent with its environment corresponds to "business architecture" as mentioned in this book.

2.2 The Concept of Business Architecture

In this book, the concept of "business architecture" is interpreted as the optimal design concept for the optimal corporate system (including stakeholders) aimed at realizing targeted business models. Established research (from an information systems [IS] research viewpoint, for example) assumes the concept of business architecture to be a design concept to integrate optimal ICT, organizational and business processes for realizing a company's deliberate corporate strategy (e.g., Versteeg and Bouwman 2006; Hasselbring 2000). From a product development research viewpoint, the concept of architecture refers to such aspects as overall product architecture, the means of division into component products through modularization, the means of assembling these components, and the mechanisms of interaction and integration among each component (e.g., Baldwin and Clark 2000).

Expanding the concept of business architecture beyond that of IT and product development strategy to a wider management perspective, business architecture means complex structures (ideal interdependencies and relationships among elements of management activity) relating to the various management activities including the relationships among stakeholders that form the target of corporate strategy action (strategy formulation and implantation); product and service development; organizational structure; business processes; and ICT strategy and other diverse management behavior. Management study researchers have categorized the definitions relating to concepts of architecture and business architecture, and the resulting concepts have not necessarily related to a comprehensive context of management. I intend that the concept of business architecture put forward here should be a comprehensive one that incorporates previous IS design and product architecture while simultaneously considering congruence with environmental change, congruence among management elements within the corporate system, and dynamic congruence on a time axis.

To optimize the corporate system, it is essential to execute modular and overall optimization of the individual management elements (strategy, organization, technology, operations, and leadership) that comprise a corporate system congruent with its environment. In this book, the corporate system architecture concepts comprising these individual management elements are referred to as "business architecture."

A company needs to achieve congruence of the corporate system (internal management elements such as strategy, organization, technology, operations, and leadership) adapting the changes in the dynamic environment (such as markets, technologies, competition and cooperation, and structures) that surrounds it. The boundaries between the environment and the corporate system (the corporate boundaries) define both the relationship with the environment and the corporate business models. Environmental change, moreover, gives rise to changes in corporate boundaries while also influencing the individual management elements within the corporate system. Conversely, the active changes of individual management elements within the corporate system modify the corporate boundaries and

Fig. 2.1 The concept of "business architecture"

further influence the environment. As I mentioned in Chap. 1, dynamic strategic management not only creates and selects (or innovates) strategies oriented to environmental change adaptation or the future, but also optimizes the individual management elements that comprise the business architecture aligned with these strategies, and corresponds to implementing total optimization as a corporate system (see Fig. 2.1).

2.2.1 Strategy as Business Architecture

As shown in Fig. 2.1, business architecture is structured from the management elements of strategy, organization, technology, operations, and leadership,[2] and congruence must be implemented appropriately among these elements integrated with the environment. Regarding strategy, environment creation and environment adaptive strategies are determined and implemented appropriately in adaptation to the environment (see Chap. 1). Companies consider both short-term profit and various future possibilities, integrate strategies targeting different areas, and implement those strategies. As the case studies of Chaps. 3 and 4 discussed, trials aimed at new business and technology innovations arising from the formation of organizational networks (cohesive networks, SWS, networked SWS, and others), including stakeholders, become important elements in realizing environment adaptive strategies. The synthesizing aspect of simultaneously implementing environment adaptive strategy, which aims for short-term profit from existing business, and environment creation strategy as a business structure common to all company targets in the case studies also receives attention.

2.2.2 Technology as Business Architecture

The business models that form the core of environment adaptive and creation strategies, the independent product and process architecture related to products and services to achieve these models, the technological innovation elements (modular, architectural, radical, and others) with the competitive power to differentiate against rivals (Henderson and Clark 1990), and business model innovation centered on marketing (Kodama 2007c) also become important. In the areas of electronics, communications devices, and machine tools, technological innovation relating to product architecture is an important theme in acquiring a sustained competitive edge (e.g., Shibata and Kodama 2008).

---

[2]I have taken up the five items of strategy, organization, technology, operations, and leadership as elements that comprise the corporate system, but I consider elements relating to people and competences, such as personnel resources and organizational capabilities, to be included in the organizational and leadership elements. Moreover, I have positioned leadership at the heart of management elements because strategy, organization, technology, and operational performance depend on the quality and style of practitioner leadership when implementing practice processes.

Moreover, the new mechanisms of the process architecture of development and goods distribution systems exploiting areas such as ICT, exemplified by the US company Dell, are indispensable for new innovators to grow. Taiwan's TSMC, for example, hitched itself to the evolution of process architecture in the semiconductor production, built standardized platforms for the semiconductor industry, and created virtual integration mechanisms that incorporated the partners (Kodama 2009). Meanwhile Sony's game business and DoCoMo's mobile phone business (described in Chaps. 3 and 4) innovations extended beyond technology to supply chain and business models, including sales and distribution systems.

2.2.3 Organization as Business Architecture

Proper organizational design is also required to realize these strategies and technologies. TSMC, for example, built networked modular organizations among stakeholders to grow the foundry semiconductor business (Kodama 2009), while Sony established its SCE subsidiary through a joint venture to build up its game business. Sony invested resources in SCE as a parent company and achieved success with PlayStation.

At DoCoMo, meanwhile, new organizations were set up in-house, operating on a project basis, to successively create business start-ups such as i-mode. The collaboration between the new and existing organizations led to the creation of new mobile phone businesses (Kodama 2007c). Leading Japanese communications device company Fujitsu also created and grew a new business as an in-house, independent venture organization by building an "ambidextrous organization.[3]" Product development through adopting new technologies took place even after the new business had separated from Fujitsu as Fanuc. Product technology conversion was achieved by building ambidextrous organizations synthesizing both existing organizations and new organizations bearing new technologies (Shibata and Kodama 2008) (see Fig. 2.2).

NTT DoCoMo, prior to the foundation, was a business division (the mobile communications division) in the advanced data communication business HQ existed within the NTT organization. This HQ comprised three business divisions (private lines, image communications, and mobile communications), which were all managed by one HQ manager. These three business divisions existed independently, but the HQ manager had a key role in the resource distribution, decision making, and personnel matters of each division. Later, government-driven communications policy led to the mobile communications division being spun off as NTT DoCoMo, which established the new mobile phone market. While the histories of NTT DoCoMo and Fanuc differ, they had points in common prior to spinoff.

---

[3] According to suggestions by Professor Tomoatsu Shibata of Kagawa University.

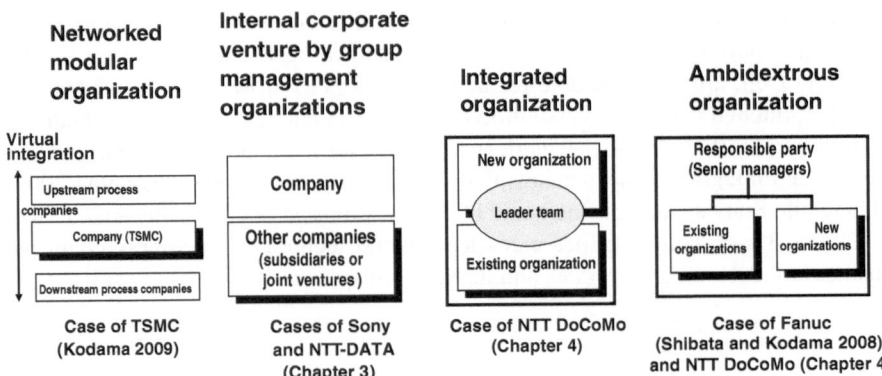

Fig. 2.2 The formation of innovative organizations: organizational congruence

Traditional, established companies emphasize costs and short-term profit, and target incremental innovation as a business model. The preferred organizational form is to implement standard organizational management under top-down, fixed mission aims, and increase production by enhancing managerial efficiency. Meanwhile new organizations face new challenges aimed at growth and innovation. Among Japanese companies of this type, middle managers ideally play a major role, displaying independent leadership and achieving new innovations.[4]

While all organizational forms have merits and demerits, the cases of Sony, NTT-DATA and Fanuc show outstanding patterns for creating businesses at speed (Fig. 2.2). These organizational forms are guaranteed appropriate support and resources (including personnel and funds) from top management. Moreover, the loose ties between the new and existing organizations mean that mutual interference or friction is rare. DoCoMo's case, however, shows a pattern where the coexistence of different corporate cultures within a single company leads to successful innovation driven by such interference and friction among organizations and personnel. The determining differences in the Sony, Fanuc, and DoCoMo cases arose from the fact that the game and machine tools businesses pioneered by Sony and Fujitsu respectively were not their main businesses at the time, but entirely separate, new businesses. Accordingly, top management could allow organizational slack (Noria and Gulati 1996; Bourgeois 1981) to implement strategies and allocate resources. At DoCoMo, meanwhile, the mobile phone business underwent a paradigm shift, and large numbers of stakeholders closely connected with essential existing businesses within the company were compelled to participate.

---

[4]Japanese companies emphasize the role of middle management (e.g., Nonaka and Takeuchi 1995). Middle management at major Japanese companies also take on the roles of driving strategy formulation and implementation aimed at cultivating new business, new product development ideas, and specific achievements (Kodama 2007a).

Ambidextrous organizations such as Fanuc are a pattern of success found with new businesses in the US (O'Reilly and Tushman 2004).[5] These US companies display ambidextrous leadership, carry the different business aims of the new and existing organizations, and simultaneously achieve short- and long-term innovation under shared visions and values. The operation of venture subsidiary operations supported by parent companies such as Sony has also been observed at major Japanese companies. In the past, Japanese companies complemented parent company businesses by establishing subsidiaries (separating the parent companies' new and existing businesses), and succeeded with synergy strategies through diversity (e.g., Rose and Ito 2005; Kodama 2007b). Organizational forms such as TSMC, which fused modular and network organizations to make networked modular organizations, are a business model at which US companies in particular excel. They are exemplified by networked modular organizations arising from virtual integration exploiting ICT, such as Dell (Kodama 2009).

Meanwhile, as explained in Chap. 4, the creation of new business by in-house integrated organizations such as DoCoMo can be thought to have the advantages of fusing corporate cultures and forming new ideas, despite the accompaniment of great friction. Markides (1998) suggests that the most important theme of strategic innovation was innovation of corporate culture. Strategic innovators at major companies structure organizations separate to the main body in order to support new strategic innovation.

Methods include building new organizations separate to the main organization (Fanuc) or establishing subsidiary companies separate to the main organization (Sony and NTT-DATA). While this pattern of organizational creation is highly effective for new strategic innovation, it is not problem-free, and brings up issues relating to the long-term fusion and harmony of new and old organizational cultures. Inevitably, the corporate cultures of parent company Sony and SCE have diverged. The greatest issue at Sony at present is how its multiple business divisions and the entire group can demonstrate strategic synergies. A key focal point in appropriate organizational design to adapt to strategy conversion accompanying this kind of business environment and technological change is how to achieve congruence of strategy and technology despite the differences arising from corporate culture and the competitive environment.

Common organizational behavior among the companies considered in these case studies goes beyond building cohesive business networks maintained by communities of practice within a company or cooperative networks established among companies to the formation of small-world structures (SWS) through cohesive network short-cuts aimed at acquiring diverse knowledge. These innovation-bent companies also feature the formation of multiple diverse SWS, bridging of context and knowledge for these SWS, and the creation of knowledge aimed at new

---

[5]By contrast, other research has indicated that general managers are a source of friction, so that interchange between the organizations in the form of a mutual general manager should be limited while interaction at a practical level should be increased (Govindarajan and Trimble 2005).

innovation. The creation of different SWS networks is a phenomenon observed in the process of merging different technologies and creating new business models crossing multiple industries (Kodama 2007a,c).

2.2.4 Operation as Business Architecture

Companies have to adopt the operations best adapted to these strategies, technologies, and organizations (building new supply chains and implementing ICT management) and build their own core competences. TSMC built its own supply chain through virtual integration exploiting ITC, SCE built a new sales and marketing system for the games business (see Chaps. 3 and 4), and DoCoMo implement regionally dispersed operations adapting to a competitive environment and real-time management exploiting ICT. Finally, Fanuc developed and introduced a production system supporting new product architecture.

2.2.5 Leadership as Business Architecture

Implementing these strategies, organizations, technologies, and operations is the task of top and middle management, who display different leadership styles in response to environmental change. A common point of focus with the case studies in Chap. 3 and 4 is the implementation of top down and middle up/down models of leadership and collaborative leadership through the formation of layered small-world structure networks (networked SWS) and typical SWS leader teams within the company. Outside the company moreover, is the pursuit of the "dialectical leadership" emphasizing relationships among stakeholders.[6] The congruence of these strategies, organizations, technologies, operations, and leaderships realizes the "business architecture" optimal for corporate growth and forms the source for creating corporate competitiveness.

2.2.6 Dynamic Process of Business Architecture

As mentioned in Chap. 4's section on DoCoMo's growth process, this business architecture is thought to change continuously and evolve dynamically (building and rebuilding) over time. Thus business architecture should not become fixed, but should have the quality of constantly changing while achieving congruence with environmental change. Through this process of change in business architecture, companies create new competitiveness and develop the potential to realize sustainable "corporate innovation streams" (see Fig. 2.3), as mentioned in Sect. 2.4. "Dynamic strategic management" is continuous, sustainable behavior aimed at creating new values. It is also a dynamic process leading to sustainable competitive dominance through the evolution of this "business architecture."

---

[6]For a general portrait of leadership styles, see Kodama (2007b), Chap. 9, and for more about Fanuc leadership research, see Shibata and Kodama (2009).

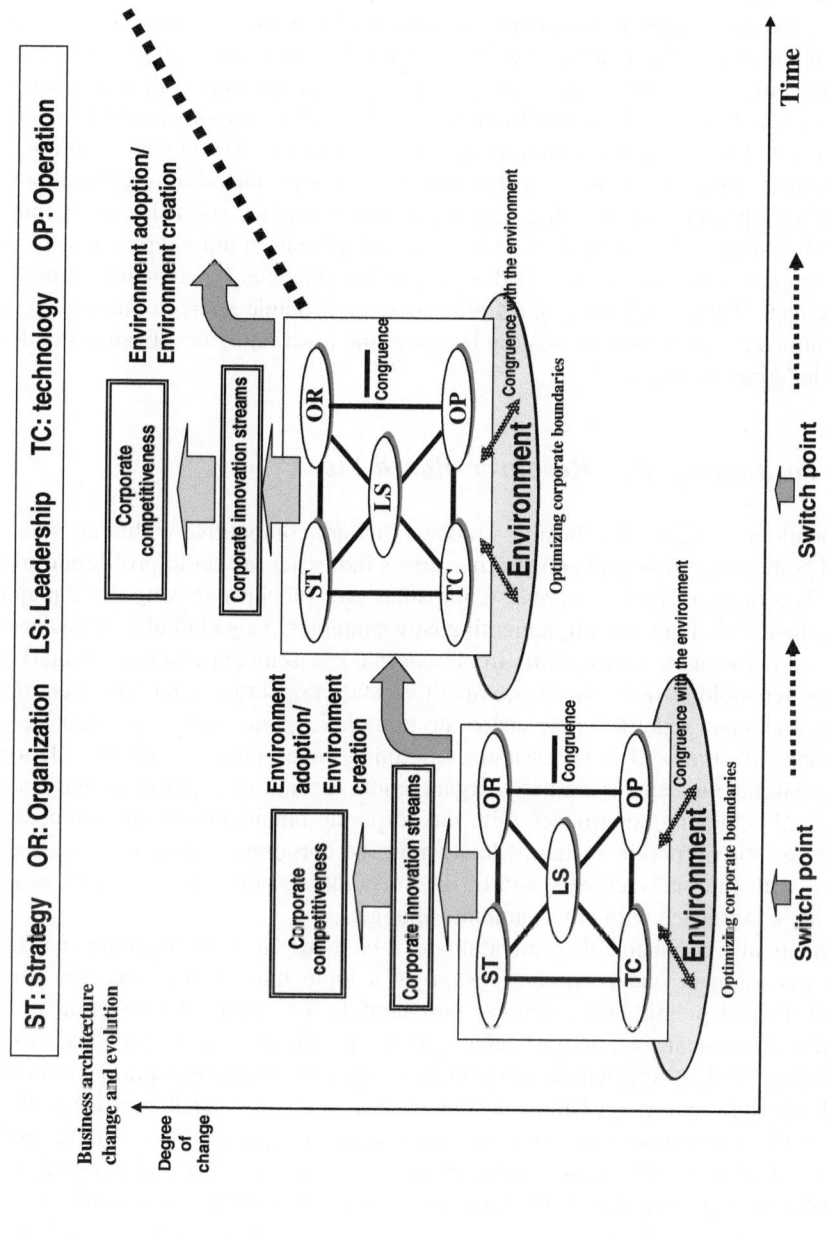

Fig. 2.3 Evolution and congruence of "business architecture"

2.3 Strategic Management in a Dynamic Environment

As mentioned in Chap. 1, the progress of ICT and digitization in recent years is increasing the need to develop products and services that merge different technologies and cross industry boundaries, and to build new business models. In these fast-changing industries practitioners must consider and act on strategies based on different competitive rules. The essence of strategic management is that companies not only create appropriate future-oriented strategies while adapting to environmental change, but build "business architecture" in line with these strategies to optimize each business element comprising the corporate system, including organizations, technology, operations, leadership, and personnel resources, and achieve continuous growth through this integrated, dynamic development. In this chapter, I will ask how awareness of issues should lead to corporate strategies for dynamic strategic management. Then I will suggest a framework for dynamic strategic management theory aimed at strengthening existing business and positioning new business under a fast-changing environment.

2.3.1 Positioning- and Resource-Based Views

As I mentioned in Sect. 1.1, business models that are competitive within an industry and both competitive and cooperative across industries are set to proliferate still further in a range of business areas. Companies are somehow sensing these major environmental changes and implementing new strategies. This kind of new business model, which leads to major shifts in an external environment comprising market changes, technology innovations, competitive and cooperative environments, and structural change, redefines new corporate boundaries and requires a switch to a new management model. It necessitates redefining the corporate system in relation to such internal elements as strategy, organization, technology, operation, and leadership, and achieving congruence with the corporate boundaries of environmental change and the corporate system. Meanwhile, the corporate system requires congruence with internal elements within the corporate system (see Fig. 1.1) while achieving congruence with environmental change.

A practically applicable theoretical model which describes the dynamic changes in the environment and corporate system is a topic that greatly concerns both researchers and practitioners. Among past models for congruence of companies and the environment, one area of interest is the positioning theory, or Porter theory (Porter 1980, 1985), which determines companies' ideal strategic positioning aims from static analysis of the external environment. This leading strategy theory provides a corporate framework for discovering an appealing position through structural analysis of the market (including analysis of competitive and transactional structures among companies). Positioning theory is an excellent competitive tool that judges a company's superiority (or deficiencies) through past, present and future (predictive) analyses of industry and competition. For the businesses mentioned above operating among different industries in a dramatically changing environment, however, the Porter theory has the demerits of being static and neglecting to analyze a company's internal resources.

A second area of interest is the resource-based view (e.g., Wernerfelt 1984; Barney 1991). This theory also describes the difference competitiveness and profitability among companies, and indicates the importance of a company's innovation processes and efficient operation resulting from intrinsic competences, resources, and capabilities. The approach of the resource-based theory is the thinking that a company's inherent internal resources determine competitive capabilities that are difficult to copy, and takes the position that a company should enter markets that exploit its inherent management resources. The resource-based view has the great merit of viewing a company's strategic position from analysis of internal resources. The major disadvantages remains, however, that it is a static theory (like Porter's), and it lacks close analysis of the external environment.

These two theories, however, provide frameworks that can be adequately applied, in the situation where grasping market structures or make projections for relatively stable environments is possible under conditions where markets and organizations can be analyzed (D'Aveni 1995 Chakravarthy 1997). Moreover, the dynamic capability approach of theories that take a dynamic strategy view are concepts of dynamic change in a company's core capabilities in line with environmental change (Teece et al. 1997, 2007). The dynamic capability approach is a path-dependent, market positioning approach whereby practitioners' thoughts and actions redefine their company's capabilities, and so strengthen market position. It is also an idea that extends from the internal (organizational view) to the external (market view), and a development of the resource-based theory.

With existing research, the positioning theory emphasizing the external environment and the resource-based theory emphasizing internal resources show two sides of the same coin (Wernerfelt 1984). As mentioned above, in dynamically changing, industry-crossing competitive environments and situations where fast-changing environments cannot be predicted, it is essential to organically link both the external environment and internal resource concepts to create a new dynamic strategic management theory. The congruences among the corporate system and environmental change and among the corporate system's internal elements, suggested in this book, indicate the framework for a dynamic strategic management theory that organically links the external environment and internal resources.

2.3.2 Environment Adaptive Strategy

The framework suggested in this book as an approach for building a dynamic strategic management theory is "change process management." Changes comprise those in the external environment and the internal corporate system, and dynamic congruence of corporate boundaries is the first proposition of these process changes. How do companies (practitioners) recognize changes in the external environment, how far and in what way should a company's corporate system be changed with regard to this, and should dynamic congruence be formed between the external environment and corporate system? To answer these questions, practitioners must interpret the degree of change in a company's external environment on a time axis, asking themselves how markets, technology, competition and cooperation, and structure are changing now and in the future, and then modify the corporate system according to the degree of change. This can be described as the congruence of external environmental change outside the company and internal elemental change of the corporate system within the company. It becomes the congruence of change in the external environment with market changes, technological innovation, the competitive and cooperative environments, and structural changes and changes within the corporate system including such areas as strategies, organizations, technologies, operations, and leadership (see Fig. 1.1).

Second is the dynamic congruence achieved among individual internal elements of the corporate system, such as strategy, organization, technology, operations, and leadership. This congruence, which comprises degrees of change in individual elements, is important. Congruence also involves questions such as how far and in what way should existing strategies be changed; how should existing organizations be reformed; how far should existing technologies be improved or radically innovated; how should existing business flow be reformed; what kind of personnel should be allocated; how and whether to cultivate or acquire; and how a congruence which resolves these issues. Despite the existence of research discussing appropriate corporate strategies when adapting to the strengths and weaknesses of environmental change (e.g., D'Aveni 1995; Chakravarthy 1997; Eisenhardt 2002), few reports deal inclusively with the factors of strategy-supported organization, operations, technology, and leadership.

Tushman and O'Reilly (1997) and Nadler and Tushman (1989) demonstrate the importance of a "congruence model of organizations" to innovation and organizational change. They posit that the basic structure of organizations comprises the four elements of critical tasks, culture, formal organizations, and people. The congruence among these blocks is aligned with the pace of innovation (the periods of incremental change and ferment), and achieved as the four elements change their substance. They indicate that the management of the conversion period from this kind of organizational congruence is a resource of sustainable competitive excellence.

According to Tushman and others, the organizational congruence model is practically important, and with regard to sudden, industry-transcending environmental change (mentioned above), it is also necessary to consider "business architecture" achieving congruence among the options and elements of strategy, technology, operations, and leadership in addition to the "organizational congruence model." Furthermore, the changes in the internal elements of these corporate systems must dynamically link to the changes in the external environment while achieving congruence. Enhancing organizational adaptability in line with the rhythm of the changes in the external environment is also required to deal with strategic uncertainty. Organizational adaptability is also the adaptability of the corporate system's strategy, organization, technology, operation, and leadership as they match the changes in the external environment.

Reinor (2007) suggests that sudden and gradual environmental change occur simultaneously, and that it is impossible to deal with the demands of an unpredictable environment, no matter how adaptable the organization may be, without some other, entirely new element. Many companies find it difficult to accurately predict situational change on a time axis from the complexities of the factors influencing external environmental change, such as market change, technological innovation, the competitive and cooperative environment, and structural change. As a corporate strategy for environmental adaptation in such situations, Reinor suggests the need to synthesize core strategies with no risk (defined as risk of failure occurring when making strategic commitments) common to multiple outstanding strategies (these can also be described as strategies whose success can be adequately projected in the planning stage) with contingency strategies contributing to a small number of, or, specifically outstanding strategies (alternative strategies with uncertain elements having the potential to help future business). As a result, companies become able to adapt to strategic uncertainty and enhance their organizational adaptability toward environmental change.

Core strategy corresponds to planned (deliberate) strategy. It also establishes a reliable competitive edge as existing business enhances core competence or related business exploits core competence. With core strategy, it is important to drive business through improvements and best practice resulting from continuous organizational learning. Core strategy should also be implemented when environmental change is slow or moderate. For business domains with high-paced environmental change, meanwhile, "strategic learning" through multiple strategy scenarios and trial and error resulting from emergent (contingency) strategies becomes important. To implement robust adaptive strategies targeting the unpredictable future, a company must enhance its environmental adaptive capabilities by implementing a range of business investments, incubations, and real-option strategies (Beinhocker 1999). The synthesis of these deliberate and emergent strategies enhances a company's environmental adaptive capabilities. In this book I will call these paradoxical strategy-making processes an "environment adaptive strategy" (see Fig. 2.4).

This environmental adaptive strategy is an objective "passive strategy behavior" that adapts to environmental change and redefines a company's strategic positioning from a positioning-based view, and that renews core competences in response to environmental change from a resource-based view. This adaptation in the face of environmental change and renewal of core competences is important. Matsushita Electric Industrial Company (now Panasonic), Canon, Sharp, Sony and others renewed their core competences in the digital consumer electronics field (including LCD and plasma large-screen TVs, DVD recorders, and digital cameras) and successively upgraded their products while expanding lineups, enabling them to maintain their dominant share of the global markets. Against a backdrop of diversity of customer needs for new digital consumer electronics emphasizing such areas as product quality, price, and functionality; technological development (especially miniaturization of devices through system LSI development, enhanced functionality, and reduced electricity consumption); changes in the competitive and cooperative environments (including cost competition arising from an offensive of South

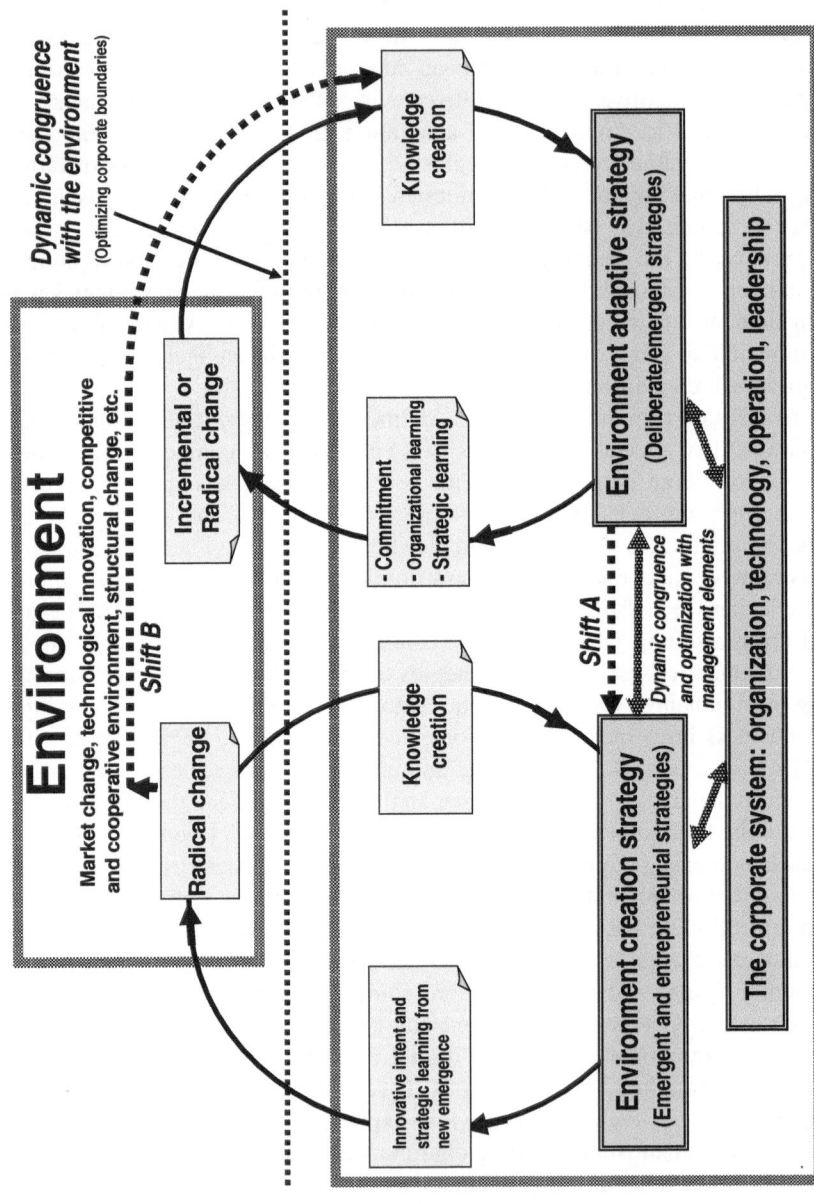

Fig. 2.4 Corporate innovation streams

Korean and Chinese companies such as Samsung Electronics, LG Electronics, and Haier, and collaborative strategies such as joint development and mergers), Japanese companies are simultaneously renewing their technological (enhancing system LSI development capabilities and introducing cell production methods aiming to cut costs and enhance quality) and operational (building supply chains including globally efficient sales, production, and support systems) capabilities as core competences while maintaining a competitive edge in their own digital electronics markets.

Among Japan's mobile phone services, NTT DoCoMo and KDDI are leading the way in adapting to the environmental changes of customer needs and technological development by continuously renewing their technological and operational capabilities and investing in new mobile phone and service markets. High-tech companies involved in the digital consumer electronics and mobile phone markets are implementing environment adaptive strategies that adapt to environmental change by continuously and dynamically modifying their core competences. This also corresponds to the dynamic capability approach mentioned above (Teece et al. 1997; Teece 2007).

Meanwhile, countless past examples of failed environment adaptive strategies exist, including the cases of failure to adapt to technological change described below. Any company hoping to grow continuously must handle the management issue of how to succeed in transferring from current to new technology. A large body of research carried out in the United States and Europe has highlighted cases of established, outstanding companies that had succeeded with current technology failing to overcome the challenges of new technologies. Thus the bearers of new technologies were not the established major corporations but the new start-ups. For example, when transferring technologies from vacuum tubes to semiconductors, which emerged in 1955 and dominated for a quarter-century, the only two companies to survive the transition were RCA and Philips (Foster 1986).

Research undertaken in the United States and Europe around this technology transfer has offered various explanations (e.g., Henderson and Clark 1990). Among these, Foster suggested the S-curve technology theory and emphasized the dilemma of technological transfer that arises from it. Christensen (1997) undertook historical analysis of the hard-disk industry and emphasized that outstanding companies failed in a factor required of truly outstanding management—that of listening to the demands of their customers. Outstanding companies able to ride the wave of new technologies certainly exist, however, and many Japanese companies, especially, have handled technology transfers successfully. Having survived the switch from vacuum to semiconductor technologies mentioned above, for example, the outstanding corporations of Toshiba, Matsushita (Panasonic), and Hitachi continue to play a leading role in technology transfer.

The recent example of technology transfer from analog silver halide cameras to digital cameras highlights a difference between US and European companies on the one hand and Japanese companies on the other. The US camera manufacturers Polaroid and Kodak have already withdrawn from the digital camera market, whereas Japanese film and camera manufacturers such as Canon, Fuji

Film, Olympus, and Nikon continue to dominate the market for both digital and conventional cameras. There is little clarification at management level as to how this difference arose or why some companies achieve so many successful strategy and technology transfers. One explanation (asserted in this chapter) is that of optimal "change process management" arising from boundaries congruence adapted to environmental change.

2.3.3 Environment Creation Strategy

In reality, while "environment adaptive strategy" forms the greater part of corporate strategy behaviour, the academic research on "active strategy behaviour" as the human subjective aspect has been much debated. Hamel (1998, 1996) mentioned strategic innovation as the key to successful value creation. Strategic innovation emphasizes the kind of business change that radically reforms (or restructures) existing business models, provides new value to the customer, anticipates competition, and creates value for all stakeholders. A corporate leader's "strategic intent" becomes an important element (Hamel and Prahalad 1989). When thinking about the corporate strategy process framework in the light of my business experience, I myself consider that the essence of corporate strategy is for a company to dynamically and mutually complement and reinforce its core competences and the constantly changing market positions that are its ideal targets while sustainably formulating and implementing strategy. Markets (external environment; positioning theory) and organizations (internal resources; resource-based theory) are not binomially opposed; rather, companies (practitioners) must possess the strategic perspective to dynamically synthesize these elements. How should practitioners think and act when opening up new markets? How should companies dynamically build and implement strategy? These questions also return to the awareness of issues mentioned in this chapter.

The strategy-making process of companies transcending their own core competences and spontaneously and deliberately formulating a series of new positions (products, services, and business models) becomes important to practitioners on a day-to-day basis, especially in an uncertain environment characterized by dramatic environmental change. To effect this process, companies must continuously work to create the new competences needed to establish newly targeted market positions and deliberately create new environments while simultaneously establishing new competitive positions through trial and error. In practice, dynamic thinking and action synthesizing an approach that simultaneously shifts from the external to the internal (corporate system perspective) environment and from the internal to the external is essential (Kodama 2007a).

What I would like to emphasize is an aspect that has rarely been the subject of academic research, that of companies (practitioners) spontaneously creating environments of new markets and technologies. Hamel (1998) asked how to create innovative strategies through drastic means and develop a strategy "S-curve." For example, the e-money services via mobile phones (mentioned above) are more in keeping with cases where companies deliberately create new markets and technologies than with those of arising from environmental changes comprising evolving customer demands and technologies. In the field of digital consumer electronics, where competition is also intensifying, electronics manufacturers such as Matsushita Electronics (Panasonic) and Sharp have also transcended individual fields of technologies and industries in an effort to deliberately create new markets and technologies. Leading examples are the ubiquitous communications market merging consumer electronics, broadband and multimedia, and the telematics market merging automobile technologies, electronics, and IT (Kodama 2007a). These cases can be seen to correspond to the S-curve of new strategies.

Accordingly, the most important issues for companies hoping to achieve future innovation is not only to adapt to environmental change but also to adopt a process of creating environments by deliberately forming new market positions with the intent of innovation through new emergence. In the high tech industry, especially, where environmental change is dramatic and market investment in new products and services essential, companies must have a dynamic view of strategy (e.g., Markides 1999) to adapt to environmental change while creating their own environmental change and acquiring new positions of competitive dominance. The dynamic strategy view, however, must go beyond the continuous creation of new products and services to establish the continuous construction of major business concepts or models. In this book, I refer to the strategy arising from the active strategy behaviour of this kind of company as "environment creation strategy" (see Fig. 2.4).

2.4 Managing Corporate Innovation Streams

Companies actualize these environment adaptive and creation strategies within the corporate system to grow existing business adapted to environmental change and realize new business aimed at creating environments, and so become able to establish a sustainable competitive edge. Figure 2.4 shows a strategy practice process framework for a company to continuously implement both an environment adaptive strategy to grow existing business and an environment creation strategy to build new market positions aimed at achieving future innovations and acquire new competences. The implementation cycle of the environment adaptive strategy adapts to a situation of environmental change (gradual or sudden progress, or a mixture of the two), and spontaneously synthesizes or separates deliberate strategy through daily commitment and organizational learning (improvement activities for daily tasks), and spontaneous strategy through strategic learning arising from trial and error with regard to strategy aims characterized by high uncertainty and hurdles. In the implementation cycle of the environment creation strategy, meanwhile, it is important for a company to implement emergent and entrepreneurial strategies (Mintzberg 1978) through strategic learning arising from trials and incubation, thus creating its own sudden environmental change through innovative intent resulting from new emergence. As described below, these two cycles are not independent but interdependent.

The implementational shift from an environment adaptive strategy to a new environment creation strategy (Shift A in Fig. 2.4) comprises action to enter completely new fields (and sometimes industries) and create new business domains through new technologies emerging in the growth processes of existing businesses. Chap. 4's case study of the domains of NTT DoCoMo's i-mode service and the IC card business (mentioned above) centered on the communications carrier and transport and retail industry correspond to this kind of strategy practice process. The case studies of NTT and NTT-DATA in Chap. 3 also correspond to this strategy practice process for a company expanding broadband services and business solutions markets through its own new strategy transfer and organizational system.

Another example of strategy practice process is the case where the implementation of an environment creation strategy results in the redefining of newly emerging business domains as a company's core business, and a company achieves a sustainable environment adaptive strategy through committing to and investing new resources in this core business (Shift B in Fig. 2.4). This corresponds to the process whereby venture companies possessing special technologies start up from an environment creation strategy, multiple rivals later enter newly emerged markets, the competitive environments of the new markets steadily change as they grow, and the expanded venture company (already grown to a medium-sized enterprise) changes gears toward an environment adaptive strategy. In Chap. 3, the subsidiaries spun out of Sony's game business and NTT's internal venture also correspond to shift B's strategy formulation process.

Moreover, in the sustainable growth of a business ecosystem (mentioned in Sect. 1.2), the renewal stage acting as the trigger creating new business becomes a key condition for sustainable coevolution with each stakeholder.[7] Here, the leader companies within the business ecosystem set out a timely environment creation strategy and promote a shift from an environment creation to an environment adaptive strategy together with the stakeholders. Then the cyclic strategy stream flowing from environment creation strategy to environment adaptive strategy and back again becomes the motive force behind the growth of a sustainable business ecosystem (Chap. 4 mentions the case of NTT DoCoMo).

With recent corporate activities, the timespan for the effective functioning of the strategy has shortened while the focus-expand-redefine cycle of strategic activity has accelerated (Zook 2007). Accordingly, companies must pioneer and promote existing business growth and new business domains while implementing a spiral of sustainable strategy practice processes by achieving dynamic congruence of environment change and the corporate system and skillfully managing environment

---

[7] Moore (1993) mentioned the progression of the following four stages of the business ecosystem development process. In the first stage ("birth"), leader companies search for new ecosystems. The second stage ("expansion") should achieve "critical mass" enveloping the customer. The third stage ("leadership") acts to achieve the shared aims of the ecosystem's main members (vendors, partners, and others), coevolves further through mutual coordination and collaboration, and reinforces a company's own strengths. The fourth stage ("renewal") renews the existing ecosystem aimed at the new targets of further ideas and innovation.

adaptive and creation strategies. In this book, I will use the term "corporate inno-vation streams" for the corporate practice process integrating these environment adaptive and environment creation strategy cycles.

Hamel mentions that twenty-first century management innovation has to enhance both operational efficiency and the strategic aspects of adaptability (Hamel 2008). This means that companies should sustainably implement the innovation streams of environment adaptive strategy. Hamel also talks about the importance of suc-cessively creating daring, rule-breaking innovations, saying "first picture the future, then create that future." Thus companies should sustainably implement the streams of environment creation strategy. In this way, "corporate innovation streams" also include and conceptualize Hamel's two assertions. Leading cases of corporate innovation streams, including Japanese automakers Toyota Motors and Honda, machine tool manufacture Fanuc, digital equipment manufacturers Matsushita Electric (Panasonic), Sony, and Canon, and communications carriers NTT, NTT-DATA, and NTT-DoCoMo also accord with these. These companies deliberately and simultaneously implement incremental innovation through sustainable improve-ment as an environment adaptive strategy and radical innovation through innovative behavior as an environment creation strategy.

Present-day business activities are complex, however, and the implementation of "corporate innovation streams" poses both a problem and a challenge. As Reinor (2007) points out, the future is uncertain, and it is unclear what strategies will succeed (cases abound where commitment to specific strategies that were sure to succeed in benefiting the whole company resulted in very little gain). This is truly a strategic paradox. However outstanding a technology appears to be, the value from the customer's point of view might be quite different. So how are cor-porate leaders and managers to handle strategy considerations? This is the main concern of this book, and one I have become aware of during many years as a practitioner.

To realize dynamic strategic management, the aforementioned concept of bound-aries congruence inside and outside the corporate system is important, and optimal business architecture relating to environmental changes and management elements such as strategy, organization, technology, operations, and leadership essential. Chapters 3 and 4 consider case analyses of Sony, NTT-Data, and NTT DoCoMo as they relate to boundaries congruence and business architecture resulting from boundary management as a corporate system. I would like to demonstrate how busi-ness architecture promotes congruence with the environment, and how it comprises a key element in sustainably implementing corporate innovation streams.

References

Baldwin, C. Y., Clark, K. B. (2000). *Design Rules, Vol. 1: The Power of Moduarity*. Cambridge, MA : MIT Press.
Barney, J. (1991) Firm resources and sustained competitive advantage. *Journal of Management*, 17(3), 99–120.

Beinhocker, E. (1999) Robust adaptive strategies. *Sloan Management Review*, 40(3), 95–106.

Bourgeois, L. J. (1981) On the measurement of organizational slack. *Academy of Management Review*, 6, 29–39.

Chakravarthy, B. (1997) A new strategy framework for coping with turbulence. *Sloan Management Review*, 38(2), 69–82.

Christensen, C. M. (1997). *The Innovator's Dilemma: When New Technologies Cause Great Firms to Fail*. Boston, MA: Harvard Business School Press.

D'Aveni, R. (1995) Coping with hypercompetition: utilizing the new 7S's framework. *Academy of Management Executive*, 9(3), 45–60.

Foster, R. (1986). Innovation: The Attacker's Advantage. New York : Summit Books.

Govindarajan, V., Trimble, C. (2005). *Ten Rules for Strategic Innovations*. Boston, MA: Harvard Business School Press.

Hamel, G. (1996) Strategy as revolution. *Harvard Business Review*, 74(6), 69–82.

Hamel, G. (1998) Strategy innovation and the quest for value. *Sloan Management Review*, 39(2), 7–14.

Hamel, G. (2008). *The Future of Management*. Boston, MA: Harvard Business School Press.

Hamel, G., Prahalad, C. K. (1989) Strategic intent. *Harvard Business Review*, 67(3), 63–76.

Hasselbring, W. (2000) Information system integration. *Communications of the ACM*, 43(6), 32–38.

Henderson, R., Clark, K. (1990) Architectural innovation: the reconfiguration of existing product technologies and the failure of established firms. *Administrative Science Quarterly*, 35(1), 9–30.

Klein, B. (1988) Vertical integration as organizational ownership: The Fisher-Body-General Motors relationship revisited. *Journal of Law and Economic Organization*, 4, 199–213.

Kodama, M. (2007a). *The Strategic Community-Based Firm*. Basingstoke: Palgrave Macmillan.

Kodama, M. (2007b). Knowledge Innovation – Strategic Management as Practice. Cheltenham: Edward Elgar Publishing.

Kodama, M. (2007c). Project-Based Organization in the Knowledge-based Society. London : Imperial College Press.

Kodama, M. (2009). *Innovation Networks in Knowledge-Based Firm – Developing ICT-Based Integrative Competences*. Cheltenham: Edward Elgar Publishing.

Kogut, B., Zander, U. (1992). Knowledge of the firm, combinative capabilities and the replication of technology. *Organization Science*, 5(2), 383–397.

Markides, C. (1998) Strategic innovation in established companies. *Sloan Management Review*, 39(3), 31–42.

Markides, C. (1999). *All the Right Moves: A Guide to Crafting Breakthrough Strategy*. Boston, MA: Harvard Business School Press.

Mintzberg, H. (1978) Patterns in strategy formation. *Management Science*, 24(4), 934–948.

Moore, J. (1993) Predators and prey: a new ecology of competition. *Harvard Business Review*, 71(3), 75–86.

Nadler, D. A., Tushman, M. L. (1989) Organizational Framebending: principles for managing reorientation. *Academy of Management Executives*, 3(3), 194–202.

Nelson, R.P., Winter, S.G. (1982). *An Evolutionary Theory of Economic Change*. Cambridge, MA: Belknap Press.

Nonaka, I., Takeuchi, H. (1995), *The Knowledge-Creating Company*. New York: Oxford University Press.

Noria, N., Gulati, R. (1996) Is slack good or bad for innovation?. *Academy of Management Journal*, 39, 1245–1264.

O'Reilly, C., III, Tushman, M. (2004). The ambidextrous organization. *Harvard Business Review*, 82, 74–82, April.

Penrose, E.T. (1959). *The Theory of the Growth of the Firm*. Wiley: New York.

Pisano, G. (1991). The governance of innovation: vertical integration and collaborative arrangements in the biotechnology industry. *Research Policy*, 20(3), 237–249.

Porter, M. (1980). *Competitive Strategy: Techniques for Analyzing Industries and Competitors*. New York: Free Press.

Porter, M. (1985). *Competitive Advantage*. New York : Free Press.

Raynor, M. (2007). *The Strategy Paradox: Why Committing to Success Leads to Failure*. New York: Broadway Business.

Rose, E., Ito, K. (2005) Widening the family circle: spin-offs in the Japanese service sector. *Long Range Planning*, 38(1), 9–26.

Santos, M., Eisenhardt, K. (2005) Organizational boundaries and theories of organization. *Organization Science*, 16(5), 491–508.

Shibata, T., Kodama, M. (2008) Managing technological transition from old to new technology: case of Fanuc's successful transition. *Business Strategy Series*, 9(4), 157–162.

Shibata, T., Kodama, M. (2009) Japanese mythology and leadership. in *Cultural Mythology and Global Leadership*. E. Kessler and D. Wong (eds). Cheltenham : Edward Elgar Publishing.

Teece, D. (2007) Explicating dynamic capabilities: the nature and microfoundations of (sustainable) enterprise performance. *Strategic Management Journal*, 28(12), 1319–1350.

Teece, D., Pisano, G., Shuen, A. (1997) Dynamic capabilities and strategic management. *Strategic Management Journal*, 18(3), 509–533.

Tushman, M. L., O'Reilly, C. A. (1997). *Winning Through Innovation*. Cambridge, MA: Harvard Business School Press.

Versteeg, G., Bouwman, H. (2006) Business architecture: a new paradigm to relate business strategy to ICT. *Information Systems Frontiers*, 8(2), 91–102.

Wernerfelt, B. (1984) A resource-based view of the firm. *Strategic Management Journal*, 5(1), 171–180.

Williamson, O. E. (1975). *Markets and Hierarchies: Analysis and Antitrust Implications*. New York : Free Press.

Zook, K. (2007). *Unstoppable: Finding Hidden Assets to Renew the Core and Fuel Profitable Growth*. Boston, MA: Harvard Business School Press.

Chapter 3
Developing New Business Models Through Dynamic Boundary Management: Case Studies of Sony and NTT-DATA

3.1 Innovation by Internal Corporate Venture (ICV)

Corporate venturing (including joint ventures) has become a key management method for strategies aimed at developing new business (e.g., Von Hippel 1997; Burgelman 1983; Block 1982; Block and MacMillan 1993; Gompers and Lerner 1999; Albrinck et al. 2001). It can be applied strategically and effectively to encourage innovation and create new capabilities as a corporate growth driver. High-tech organizations now exploit corporate venture collaboration, both inside and outside the corporations, as a common way of dealing with today's complex business models. With ICT business, especially, close collaboration among actors from different backgrounds and industries is a crucial factor in creating new knowledge and business models. The new business ideas can be produced through the formation of organizational networks, including external partners and customers (e.g., Kodama 1999).

This chapter analyzes frameworks leading to the success of internal corporate ventures (ICVs) from the perspective of business model innovation, focusing on the evolutionary process of dynamic strategic management through ICVs connected with major corporations. Established corporations form ICVs as new, highly strategic businesses created with a specified degree of independence from the main company's business domains. ICVs can exploit the large corporations' rich management resources while possessing the venture company strengths of superior speed and manoeuvrability. ICVs can also exploit the management resources of the group companies (including the parent company). The consolidated management ideas of the group enable bold business development, facilitating entry into markets that the main corporation would find difficult to attack. The accumulation of the companies' heuristic experiences of success triggers a transformation in the culture of the major corporation away from safety toward challenge, while linking to the new dynamism acting as a stimulus within the corporation.

As mentioned in Chap. 2, knowledge boundaries arising from the differences in philosophies, specializations, own "thought worlds," and "mental models" exist among practitioners at the boundaries of ICV-type mixed organizational bodies, cross-functional, cross-organizational teams, and organizations distributed inside

M. Kodama, *Boundary Management*, DOI 10.1007/978-3-642-03789-4_3,
© Springer-Verlag Berlin Heidelberg 2010

and outside the company. By forming business models, practitioners come to perceive and recognize a large number of knowledge boundaries. The presence of these "knowledge boundaries" can simultaneously restrict innovation and act as a trigger for new knowledge creation. The presence of the small-world structures (SWS) or small-world networks (SWN) formed inside and outside the organization (described in this chapter) accelerates innovation with these knowledge boundaries. SWS (SWN) provide a bridge (short-cuts and rewiring) among formal organizations possessing features as cohesive networks, and become enablers aimed at new knowledge creation.

This chapter examines new business development at Sony and NTT-DATA as ICV case studies. It analyzes the organizational behaviour of practitioners forming leader teams and SWS networks as small-world structures that accelerate development of new business models and product innovation for the venture businesses of major corporations. Then it suggests new insights into how the practitioners' practice processes of SWS and SWS network creation promote the gathering and integration of distinct creative and practical knowledge, and achieve congruence between the environment and corporate system and among individual management elements within the corporate system.

As with Sony's game business and NTT-DATA's ICT business, I believe that clarifying the mechanisms of knowledge integration, which transcends organizational boundaries inside and outside the company and accompanies business model complexity, will demonstrate its beneficial implications (both practical and academic) to a large number or practitioners. In the case studies presented in this chapter, practitioners commit to numerous knowledge and organizational boundaries, bridge multiple knowledge boundaries possessing distinct contexts and knowledge through the formation of SWS (networked SWS), and achieve knowledge integration. Their aim is to go beyond the development of products and services to realize complex business models incorporating new supply chains. Network concepts crossing practitioners' knowledge boundaries become enablers creating new business models.

3.2 Sony: A Case Study

3.2.1 Starting Up a New Business

In December 1994, Sony group company Sony Computer Entertainment (SCE) released PlayStation, a household games device loaded with leading-edge system LSI games that used innovative architecture to process ground-breaking, three-dimensional computer graphics images. It took just three years for SCE to overhaul Nintendo's dominance of the games market and transform the business model of the games industry. Sony's strategy for dominating the games device market includes successful product development and a close connection between innovative marketing strategy and technical strategy. Viewed from an organizational behaviour perspective, the method of trial and error combined with hard work led innovative actors in diverse specialized fields, which transcended industrial, organizational, and knowledge boundaries, to discover and successfully implement a new business concept.

The newly established SCE assembled an array of human resources. From Sony itself came Ken Kutaragi's PlayStation development team and core members of Sony Laboratory. Teruhisa Tokunaka (first president of SCE), a Sony management strategy staff member, and other business and financial management staff also arrived for the essential unification of SCE's business strategy. The games software marketing strategy and sales team came from SME in the form of Shigeo Maruyama (now SME president and a director of SCE), who was responsible for Epic Sony Business Division, his New Media Division (development and marketing of PCs and Famicom [family computer] games software) subordinate Yuji Takahashi (later SCE director and general manager), and his team. These prominent professionals were joined by Akira Sato (now an SME director), who came across from CBS and Sony Records with marketing and sales expertise.

As a late starter with dominant players Nintendo and Sega Enterprises already present, SCE's new business concepts revolved around what they were capable of doing and what they had to do. Their work began with a thorough study of Nintendo's business model. They listened to the no-nonsense voices of players and customers in the Nintendo setup, including software makers, wholesalers, retailers and users, and gathered a range of problems and complaints. Then, armed with this raw data, they turned their attention to defeating Nintendo.

What kind of business model could put the user at the top and provide all players with reasonable benefits while keeping the user happy? SCE's basic plan was to eliminate the dissatisfactions of software makers, retailers, and users garnered from a factual analysis of Nintendo's game business, thus reversing the weaknesses of Nintendo's business. Achieving this meant building the value chain of a new business model involving SCE cooperating with software makers and retailers in a win-win situation.

At SCE, the assembled core members began discussions aimed at strategic decision-making. They expressed their views over a platform to support the games software makers and the means of building a win-win relationship among all players, including end users and retailers. First the members discussed technical strategies to provide a platform (formats and other areas) for an attractive games device that would encourage software creators to design games software. Then they considered software strategy to show whether there was a business structure (incorporating licenses and royalties) which would attract the software makers' managerial ranks to develop and provide software. Finally they talked about a distribution strategy to build a distribution system that would provide all players with a win-win situation.

Kutaragi's development team was the prime mover on the first technical strategy item. The team conceived the basic architecture of an LSI system integrating the world's leading three dimensional computer graphics. At that time, the design of the specific logic circuits for this LSI and their layout and construction were developed jointly with the US-based LSI Logic Company, which was the world leader for LSI at this level. The team came up with a device to be mounted in an ordinary household games console that would rival the functions of the US-based Silicon Graphics Company's computer graphics work stations (priced at more than ten million yen each). An order for one million units from SCE provided the economies of scale that enabled it to be priced for an ordinary household. Kutaragi also predicted the trend towards lower prices and the increased functionality of components such as semiconductor memory, and he planned from the outset that LSI architecture development should enable component numbers to be reduced in stages. These measures improved production, enabling the hardware to be mass produced, and this subsequently led to lower prices for the PlayStation.

The second technical strategy item involved the use of CD-ROM optical discs as media for the games device instead of the mask ROMs that had been in use up to that point on the Famicom. The CD-ROM was best suited to handling the high volumes of data for the pictures, sound, and moving images of 3D games. Kutaragi's team compensated for the CD-ROM's weaknesses of poor access time and operability with a means of drawing in real time that reads all CD data into the games device together and uses an image and compilation engine that had been developed.

The third technical strategy item involved the provision of software support tools for games software creators. This comprised a range of libraries for the on-line support of creators and their software works. Kutaragi's team then perfected a user-friendly environment for the creators to make their software.

Software strategies mostly aimed to appeal to the management divisions of the software makers. One such strategy closely resembled the technical strategy that exploited CD-ROMs. The members of Maruyama's and Takahashi's Business Division proposed new royalties for software makers based on the use of the CD-ROM, and joined Kutaragi's team in promoting the involvement of the software makers. The CD-ROM is not only better suited to holding larger volumes of multimedia data than the mask ROM, but also has the great advantage of low production costs. The software makers discovered they could cut production costs by two-thirds if they used the CD-ROM. Meanwhile, software OEM production fees and changing attitudes toward royalties brought about dramatic reform of existing Nintendo-based rules.

As a result, the CD-ROM manufacturing fee (including commission production and royalty fees) paid to SCE by the software makers was set at 900 yen. This compared favourably with the 3,000 yen fee for the Nintendo mask ROM system, and represented a considerable cost saving for the software makers. The CD-ROM also allowed additional repeat production, eliminating the need for the 500 yen risk-avoidance fee levied due to read errors that occur with the mask ROM. Finally, CD-ROM production can be increased within three working days, making it possible to construct a software supply chain obviating the need for dead stock storage costs. To create such a supply chain, SCE arranged a factory equipped with the latest facilities, including an enormous delivery center at SME's subsidiary JARED (record delivery) Shizuoka and the latest CD-ROM press lines. These business strategies dramatically reduced the hurdles for software makers participating in PlayStation, and made the business side more attractive for managers.

The third item is a win-win distribution system to benefit all players. SCE has implemented a distribution policy that is revolutionary for the games business. Sato's sales promotion team exploited its experience and knowledge of the record business model (especially the area of responding to user demand) to bring it over to the games business. The music business basically involves low-volume production of numerous items, and the music CD business model involves collecting complete sets of multiple titles and selectively purchasing works from the sets in accordance with user preferences. The supply chain can promptly deliver higher volume as soon as a work becomes a hit. Since repeat supplies can readily be arranged, there is no need to stock CDs, and a suitable distribution system supports direct purchasing by SCE and direct selling.

The beauty of this system is that software that sells can immediately be reproduced, production and inventory are balanced, and problems associated with the former mask ROM, such as package deal selling, wholesaling, and pressuring of retailers, simply do not occur. SCE had achieved unified marketing, establishing a system of specialization in which the software makers promote sales and SCE does the actual selling. Since SCE knows accurately and in real-time the number of games devices shipped and the amounts of software shipped and supplemented, it is able to build appropriate supply chains while exercising unitary control over marketing data, giving feedback to the software makers on predictions of future demand and data for the development of new software.

This direct purchasing system also aids the strategy of increasing software-maker participation in PlayStation, especially planning participation by small-to-medium software makers with limited funds. The system is also a device for cultivating potential skilled software creators and having good products created by software makers who lack the capacity to sell software themselves. Nintendo exercised unitary control over software from content through production and quality. It believed that a market could be taken with a few interesting software products, and several Nintendo managers believed the games market would be destroyed by bad software. SCE, on the other hand, stressed the potential capabilities of software makers and freedom of expression for software developers. Consequently, SCE left its door wide open to those who wished to make games software. It found excellent creators and endeavored to discover outstanding talent for the next generation from those who designed its diverse range of software products. This is analogous to the idea that a mountain will naturally rise higher if the range of its foothills expands.

The mutual potential implicit in the technical, software, and distribution strategies outlined above led SCE to draw up license contracts with more than 200 companies, including Namco, Konami, and Square, by August 1994 (before PlayStation was released). Contracts were also drawn up with 50 corporations and 3,000 bulk and retail stores, and a sales network was steadily built up incorporating major toy and record wholesaler SME and wholesalers for electrical shops in the nationwide Sony Group. Then in December 1994 PlayStation was introduced, selling 150,000 units on the opening day. The PlayStation Series (PlayStation, PlayStation2, PSX,

PlayStation Portable, PlayStation3) is now distributed throughout the world. Its success derives from the synergy of the three strategies described above. PlayStation is an excellent platform that appeals to software makers and provides an attractive environment for creators. The user-friendly environment boosts the motivation of the software creators, who create large numbers of programs. The effect is to suddenly increase the volume of software works and games device hardware shipped, leading to a rise in PlayStation user numbers. In technical terms, it generates positive feedback from network externalities (Shapiro and Varian 1998). This PlayStation-centered games business resembles the model for a mobile telephone service that can connect to the Internet. As Chap. 4 explains in detail, NTT DoCoMo i-mode service, which provided the world's first cell-phone enabled Internet service, led to greater content for mobile phones, more subscribers, and positive feedback in terms of increased phone shipments (e.g., Kodama 2002).

3.2.2 Integrating Knowledge from Different Fields Through Network Concepts

These PlayStation business models integrate knowledge of hardware from games devices, software from games, intellectual property from copyrights, royalties and other sources, and marketing and management from distribution and sales channels. When SCE was established, staff members Kutaragi and Tokuyama from Sony and Maruyama, Sato, and Takahashi from SME, together with their teams, opposed and confronted each other as they searched for a new games business model through trial and error. The Sony side knew about consumer electronics, computer technology, and business models, while the SME (originally a record company) side looked after the development and marketing of games software for Nintendo. Together with Epic Sony members, who had games business experience, they had come together to form the single organization called SCE. Not all the members at that time knew in detail the commercial customs of the games world headed by Nintendo. However, they studied the existing persistent business structure and ascertained the problems and issues. Then as SCE, they proceeded to hammer out new business strategies that would creatively destroy the existing business model. What was the driving force that supported the actions of these SCE actors?

Maruyama, who was in charge of games at SME's Epic Sony, was very interested in games software and keen to write his own groundbreaking game at some point, but did not have the hardware to achieve this. Meanwhile Sony's Kutaragi was intent on developing a games device that would achieve three-dimensional computer graphics. He had the technical ability to do this but was unable to write the software. So Murayama joined through the medium of games, while Kutaragi contributed enthusiasm and faith. Norio Ohga, then president of Sony, had discovered Kutaragi's talent earlier on, and had put Kutaragi and Maruyama in the same office before establishing SCE. Thus Kutaragi's and Maruyama's teams shared the context surrounding the future games business at an early stage, giving rise to new meanings. Tokunaka, a business management professional, and Sato, a record business professional, together with their teams, also shared that context. Each of them debated a wide range of ideas, and a definitively coherent context was formed and shared. They queried the reason for the existence of this new context on behalf of SCE, and displayed the will and strength of purpose as actors to realize this precise strategic objective: "To radically reform the games business with our own hands!" (Bruch and Ghoshal 2004).

Individuals demonstrate leadership, but specific individuals are not there to persuade a team, nor to coerce team members. The starting point for leadership behavior in SCE is collaborative leadership (Chrislip and Larson 1994; Bryson and Crosby 1992) in which individuals acknowledge each other's abilities, complement their partners' abilities, have their own abilities complemented by their partners, and collaborate. This creates leadership teams (LT) as a small-world structure (SWS) in Fig. 3.1.

Successful factors in project management include setting a common strategy objective among members and having the project leaders take a central role in establishing a stage to achieve this objective. This stage should create an environment that draws the power of all members and enables project members to act harmoniously and autonomously. The central role of this "setting the stage for power" has been one of the factors behind the success of the PlayStation business. "Setting the stage," for example, also provided a connection between Kutaragi's development team and the software maker creators. Software development tools and online support were given fully and unsparingly to support the creators' innovative actions and enhance their motivation. SCE joined with creators throughout the world to provide end users with the new value of "computer entertainment," and this new value became the great driving force behind PlayStation.

3.2.3 SWS and SWS Network Creation

Innovation stems from conflict among heterogeneous fields of technology (Leonard-Barton 1995). Surely this PlayStation business concept was also born from conflict between actors within SCE whose backgrounds were in heterogeneous fields of technology. New knowledge was formulated by the skilful blending and integration of the knowledge possessed by individuals within SCE: Kutaragi's superior technical power, the creative and imaginative conceptualization of Maruyama and Takahashi, Sato's record business marketing know-how, and the holistic business planning faculty of Tokunaga at the top. The various repetitive conflicts and frictions among individuals have given rise to new knowledge in the form of definite strategic objectives.

A feature of SCE is that the task of building a strategy for a "win-win relationship for all players, including users" gave rise to creative discussion and collaboration. The effect of discussion was not to compromise but to build a dialectically strategic objective while reiterating dissent, and the point is that they acted toward

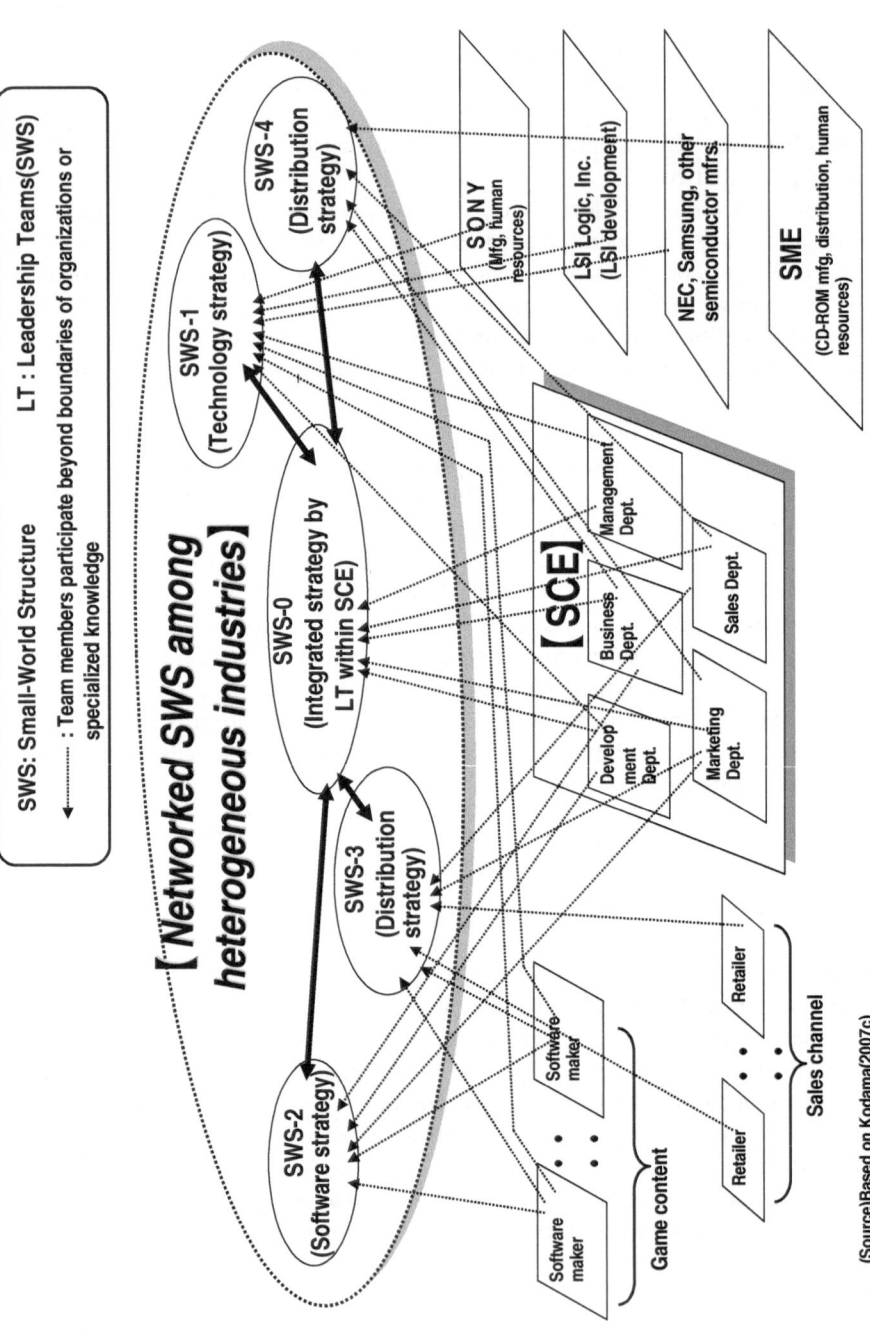

Fig. 3.1 Networked SWS for nurturing playstation business

achieving this (Kodama 2004). In Fig. 3.1, strategic objectives that anyone can understand and accept dialectically are formed in the leadership teams (LT) within SCE. Accordingly, resonance (Kodama 2001) and confidence (Vangen and Huxham 2003) in mutual values become the basis and driving force for achieving strategic objectives.

Collaboration with various players was key to the positive implementation of technical strategy. Developing PlayStation (the games machine with leading edge technology) involved making tight, cooperative connections with semiconductor makers such as LSI Logic, Samsung Electronics, and NEC. Sony's Kisarazu factory was fully exploited to assemble the sets. PlayStation's features were not confined to functional aspects; Sony's designers were also seeking to create a unique and original games device incorporating a controller. The design aspect is high on the list of factors that cause a product to become a hit with consumers (the mobile phone is a typical example of this). However, a complex design that is difficult to mass produce would mean that product supply could not meet demand, and opportunities to secure users would be lost. Therefore Sony's designers asked themselves "What kind of design can be mass produced?"

The designers came together with the engineers responsible for production and metal mold-making at the plant to obtain a definitive solution to the problem of compatibility between product design and improved productivity. The result was a novel design for the PlayStation body and controller that helped make it a hit. Among the software makers, moreover, the prototype games machines and range of development tools functioned as boundary objects (Star 1989; Cramton 2001), and the strong sharing of knowledge between Kutaragi's development team and the creators gave rise to knowledge inspiration that led to the making of a better product.

In implementing these technical strategies, strategic alliances were formed with a number of enterprises. SCE and other companies shared their knowledge closely. Facing the strategic technical objectives together inspired them to share mutual knowledge and accelerated the creation of new knowledge. New objectives were faced in the course of repeated business meetings with other companies, and these confirmed the levels of affirmation and attainment. Many small-world structures (SWS) came to be formed in space-time intervals from the strong sharing and creation of knowledge between these companies (SWS-1 in Fig. 3.1).

The software strategy involved action to appeal to software makers thinking of participating in the PlayStation format. Before PlayStation prototypes became available, hardly any software makers participated, as they doubted the feasibility of games devices with three-dimensional computer graphics functions for household use, but SCE leadership teams (LTs) persisted with explanations. After a demonstration meeting with a PlayStation prototype convinced software makers of the PlayStation's viability, however, the SCE leader team got the software makers actively involved. Having discovered common values in the cost structure of games software that exploited the benefits of PlayStation performance with the CD-ROM, the software makers participated enthusiastically in the creation of new software products. SCE built numerous networks with software makers, and these networks formed ToBs at space-time intervals for the close sharing and creation of knowledge (SWS-2 in Fig. 3.1).

The distribution strategy involved SCE calling on software makers and retailers and persistently explaining the new distribution setup with its bundled buying system and prices. Many software makers were opposed to the bundled buying system, in particular, but gradually came to understand that SCE had a higher vision to "achieve optimal supply chain management." This distribution system was the first to be trialed in the games industry. Software makers familiar with Nintendo's existing commercial customs found SCE's system difficult to understand at first, but the opportunity to propose prices (the ability to procure new software inexpensively and promptly), not just for their own companies but for users who had bought software, made them change their minds. As with the technical and software strategies, the distribution strategy involved forming strong networks with many software makers and retailers. These networks formed numerous SWS for the close sharing and creation of knowledge at space-time intervals (SWS-3 and 4 in Fig. 3.1).

These various SWS, which are dissimilar in context and knowledge and transcend different businesses, have come to be strongly linked to LTs within SCE. The linkage of SWS allows different contexts and knowledge to be shared dynamically among various organizations and actors, and action taken on new problems and issues. This network of LTs and SWS is neither physical nor visible, but comprises invisible networks deliberately formed by actors. The driving force of the PlayStation business came from the project network of LT and SWS networks crossing the company's internal and external boundaries, and a new value chain constructed in the games business. This project network transcending the company's boundaries is dispersed within each SWS and integrates knowledge implemented by the three strategies.

3.3 NTT-DATA: A Case Study

3.3.1 New Business Model Development

At the beginning of 1999, NTT-DATA CEO Dr. Masaharu Aoki felt the need to add delivery of ICT services to the company's role as a system integrator manufacturer. System integration (SI) has been a core competence of the NTT-DATA group for many years, and will continue to be a source of competitive capability in the future. The SI business, which comprises around 80 percent of NTT-DATA group sales, builds information systems in response to customer needs. With technology innovations in fields such as ICT for future broadband expansion and with various changes in the political, economic, and social environment, the NTT-DATA group customer companies came to provide new products and services. Dr. Aoki keenly recognized the need to respond swiftly to the market changes.

NTT-DATA saw its transformation into an ICT service supplier as an urgent task. This transformation included building and providing systems in response to client company needs, pioneering environmental change, and providing businesses and services with the potential to exploit ICT. It also involved working alongside the client company to develop new business and services using ICT.

Broadband penetration had an especially great impact on the NTT-DATA group business environment. The spread of broadband involves network development, and for some products the day has already arrived when networks simultaneously link consumers, producers, and raw material producers, or related businesses such as transport, insurance, and settlement. The mass production and consumption era is giving way to an industry structure that maintains an environment where new businesses arise through broadband, and individuals access what they want at a reasonable price on broadband networks.

NTT-DATA predicted such trends, and promoted the collaboration strategy that comprises the core framework for many of the client companies mentioned in this book. NTT-DATA maintained and developed conventional SI competences as a system integrator, promoted close collaboration with client companies who demonstrated an ambition for new businesses as an ICT partner, and promoted the development of platforms, including billing, settlement, and security, that create new e-business as a service provider.

3.3.2 Building Flattened Organizations and Organic Networks

NTT-DATA transformed its organization to build new bodies aimed at realizing new strategies, including separate public, financial, and industrial business headquarters promoting conventional SI business. Thus in September 2000, the business planning and development HQ and regional HQs for information network businesses were established to promote the new ICT partner and service provider services. The main missions of the business planning and development HQ were as follows.

- *Launch IT partner businesses.* Jointly create new businesses and services (new business models) with other companies around an IT core.
- *Promote IT utility businesses.* Plan and develop common and basic services as IT core businesses.
- *Cultivate customer relations aimed at new SI acquisition.* Accept SI orders from new joint ventures and for joint venture partners' mainstay businesses.

The mission of the information network business HQ in realizing new business through technology focused on services for settlement-related networks, domestic credit business outsourcing, finance and card-type processing, and corporate network systems.

In this way, NTT-DATA's integrated organizational innovations established an HQ for business planning and development and corporate HQs for information network businesses to cultivate new business in parallel with traditional organizational bodies produced by conventional SI business. NTT-DATA also promoted SI business as an exploitative activity of current business while co-establishing and pursuing new business as exploratory activities for future business cultivation.

Figure 3.2 shows NTT-DATA's internal organizational structure, comprising the new organizational bodies (business planning & development HQ and information network business corporate HQ) of flattened formal organizations and the SI business HQs of traditional organizations. To realize speedy, mobile management through mutual collaboration, NTT-DATA avoided bureaucratically layered structures, established a total of 73 business units (corporate HQs) comprising project-based organizations (e.g., Kodama 2007c), and achieved a flat organizational structure that arranged these business units in parallel. NTT-DATA built in-house networked SWS as internal organizational networks directly controlled through management meetings attended by the CEO, and built collaborative networks spanning individual organizations. NTT-DATA's model of internal organizational structure co-established exploitative and exploratory activities as cooperative strategy. It became a source for the creation of organizational capability conferring a sustained competitive edge. These competences arose through collaborating with other companies on the processes of developing, introducing, and applying ICT.

Meanwhile, it became necessary to build both optimal internal organizational structures and new external organizational structures. A key task for NTT-DATA was to co-create new business models and expand horizontal boundaries through close collaboration with companies spanning industry boundaries. As mentioned in Chap. 2, in order for companies to optimize "congruence with the environment,"

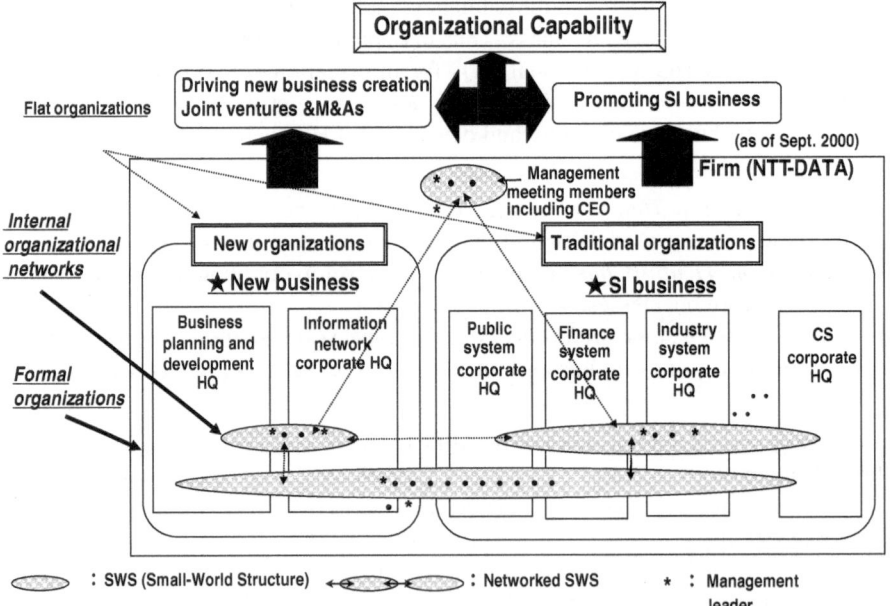

Fig. 3.2 NTT-DATA's internal organizational structure

the vertical and horizontal boundaries must be skillfully rebuilt. NTT-DATA applied and adapted corporate ICT resources in diverse sectors essential to building value chain structures (including sales, production, and procurement; client and after-sales services; business management; and ICT management) consisting of vertical boundaries around the axis of traditional core competences (SI competences). The company also expanded the horizontal boundaries of its own business domains.

Examples of the major strategy goals that emerged are the application, in various corporate domains, of next-generation transaction management systems (SCM solutions) to seamlessly link the business processes of sales, production, and procurement among companies; next-generation client management systems (including CRM and billing solutions) supporting the business processes of clients and after-care services; and ERP/HCM (Human Capital Management) solutions supporting new corporate governance, considering areas including group management for business planning and internal systems. The expansion of NTT-DATA's horizontal boundaries and building of individual value chains (vertical boundaries) applying new ICT in individual corporate domains were also major strategy goals.

Thus NTT-DATA acted to establish new joint venture companies, and accomplished this through strategic alliances with external partners in different industrial sectors. NTT-DATA then aimed to optimize "congruence with the environment" and "congruence within the corporate system" aimed at co-creating new business with numerous external partners, and built new external organizational networks (see Fig. 3.3). The creation of group companies through joint ventures developed as a

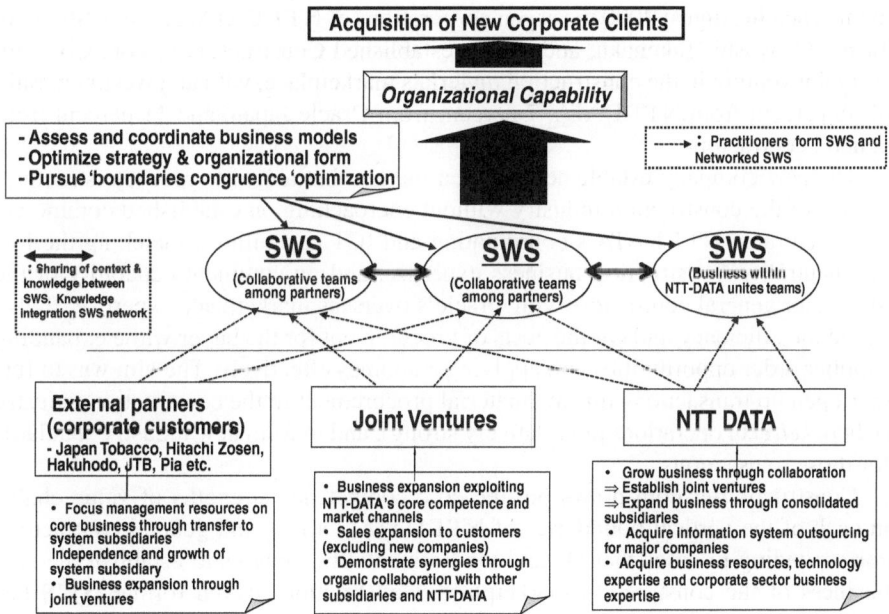

Fig. 3.3 NTT-DATA's external organizational structure

result of 'SWS' formation. NTT-DATA established joint-venture corporate groups as formal organizations by creating external organizational networks of networked SWS with client companies, and acquired and accumulated competences required for new e-business. Next, NTT-DATA built win-win relationships and targeted co-evolution of the whole sector by assessing and coordinating business models through building external organizational networks of networked SWS among external partners.

Below I will analyze cases of innovation in construction, marketing, and content resulting from NTT-DATA's strategic alliances and joint venture strategy.

3.3.3 Construction Innovations

At the start of the twenty-first century, Japan's construction industry market comprised around 570,000 companies, and investment in construction amounted to some 70 trillion yen, of which a huge 28 trillion yen (40 percent of turnover) comprised construction equipment. Recently the need for accountability to clarify costs has risen in such areas. The construction industry features numerous material and machinery contracts and a high distribution volume, factors that make it suited to web business, and ICT application is expected to take the lead in enhancing production. With this background and the latitude to boost productivity using ICT, construction industry companies investigated linking up with NTT-DATA or Oracle to enter the new marketplace of intermediary e-commerce for construction equipment. Then in August 2000 the seven corporations of NTT-DATA, Kajima, Shimizu, Taisei, Obayashi, Takenaka, and Oracle established Construction-ec.com Co., Ltd. as a joint venture in the construction materials marketplace, with an investment ratio of 38 percent from NTT-DATA, 7 percent from Oracle Japan, and 11 percent from general contractors.

The new company established an open market that aimed to promote structural reform of the construction industry without encroaching on established commerce. It concentrated NTT-DATA's coordinating and ICT capabilities, matching the formal neutrality, construction business expertise, and procurement capability of the five major general contractors with Oracle's overseas marketplace expertise. It also raised the efficiency and cut the costs of procurement for the buyer while expanding supplier order opportunities and applying resources effectively. The aim was to further open up transactions for raw material procurement in the construction industry (where *keiretsu* operations are relatively strong), and to form new industry standards for those transactions.

Construction-ec.com draws out the core competence strengths of value chains in each of the vertical boundaries of NTT-DATA, Oracle, and general contractors, corresponding to cases that expand the horizontal boundaries to establish the new business of the construction marketplace. Construction-ec.com formed SWS targeting collaboration among the NTT-DATA and Oracle platform innovators and the five general contractors' innovators, and planned to optimize the 'boundaries

congruence' (congruence with the environment and congruence within the corporate system) by assessing and coordinating the new business model. Finally,
Construction-ec.com was established as a new joint venture (see Fig. 3.4). The alignment of Japan's five general contractors had great significance. The new company
appealed to the participation of general contractors, subcontractors, and manufacturers as well as the investing companies, and aimed to create a new standard for
construction business transactions.

Construction-ec.com operates through virtual integration by forming organizational networks exploiting ICT in business areas ranging from estimate requests for
lease rental of provisional construction materials and equipment to orders, billing,
and payment. The e-commerce marketplace established by Construction-ec.com
also hosts estimate requests, orders, distribution, and settlements among buyers and
sellers. Services to supply surplus construction machinery and materials, provide
information on second-hand goods, arrange distribution of goods, and act as a debt
recovery and joint purchasing and procurement agent were developed through ICT-
centered virtual integration. Buyers became able to cut the costs of business and
negotiating processes for procurement, and suppliers to expand order opportunities,
cut transaction costs, and effectively exploit unused assets.

Expanding the objectives for construction machinery and materials and providing information and ASP services has also contributed to "construction portal" type
business development offering daily access to construction business and enhancing
promotion of ICT applications in the construction industry. The portal features an

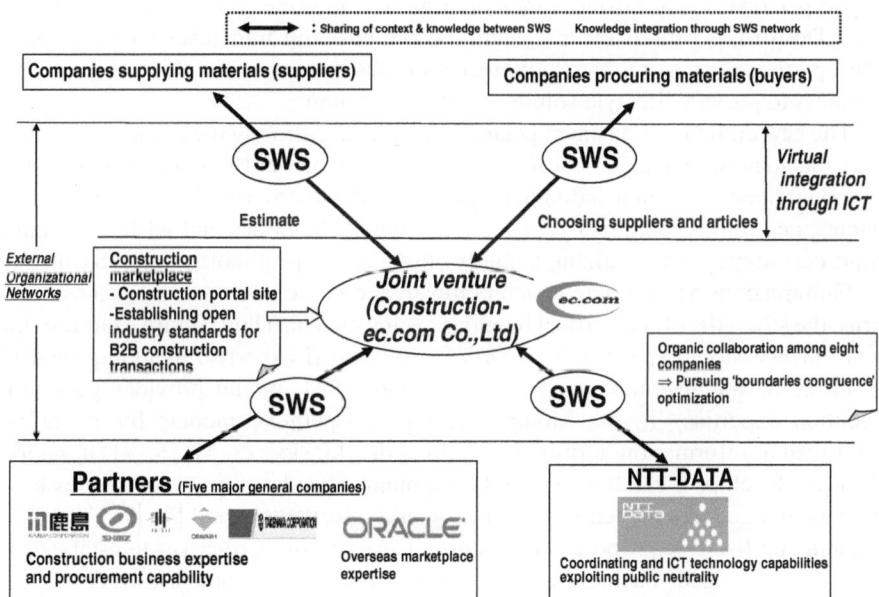

Fig. 3.4 Organizational networks centered on *Construction-ec.com*

advanced B-to-B e-commerce market built around the alliance of the five major general contractors. The site promotes services with detailed knowledge of construction industry procurement processes to enable a company to search for ideal business formations without encroaching on existing business, and to plan and make proposals for efficient and effective business. It applies the latest ICT technologies appropriate to business progress, and targets open industry standards for construction transactions. To put its plan into action, Construction-ec.com guaranteed integration with CI-Net, the construction industry's EDI standard.

3.3.4 Marketing Innovations

In May 2000, Family Mart combined with Itochu, NTT-DATA, Toyota Motors, Dai Nippon Printing, JTB, and Pia to establish a joint venture named Famima.com, which was to build a system combining EC and convenience store (CVS) franchises. The investment ratios were 50.5 percent for Family Mart, 14.5 percent for Itochu, 10 percent for NTT-DATA and Toyota, and 5 percent each for Dai Nippon Printing, JTB, and Pia. The new company was established with the goal of promoting a personal agent business to guide consumers. Famima.com established the consumer angle from the concept of guidance for the supply side of conventional business products and services. The result was the personal agent business, with services providing mechanisms giving hotel concierge-type support for the consumer-side processes of purchasing, information acquisition, and selection. Other examples of the personal agent business include Yahoo!, Priceline.com, Amazon, com, and NTT DoCoMo's i-mode service. Famima.com has built franchises with the Family Mart participating stores as real e-business bases and promoted this personal agent business to provide lifestyle solutions for the consumer.

The new company's business goals involved concentrating the participant companies' outstanding management resources on Famima.com by building partnerships of top companies in their individual fields around the criteria of the investing companies, and maximizing that capability to support the hopes and wishes of Family Mart customers while realizing a highly effective and profitable e-retail business.

Famima.com has extracted each company's core competences, and these comprise the strengths of the vertical boundary value chains. Thus Family Mart provides convenience store expertise, NTT-DATA provides SI expertise including total EC (such as IC membership), customer data, and CRM, Itochu provides goods distribution expertise, Toyota Motors develops cooperative models for in-car and multimedia information terminals (multimedia kiosks, or MMK), Dai Nippon Printing develops e-catalogues and data content, JTB develops and provides travel content and delivers related product lineup and information, and Pia provides ticket content and lifestyle proposal information. Famima.com's role was to establish the new personal agent business and expand the horizontal boundaries. NTT-DATA, Family Mart, Itochu, Toyota, Dai Nippon Printing, JTB, and Pia formed a seven-company SWS with the aim of collaboration, and aimed to optimize "boundaries

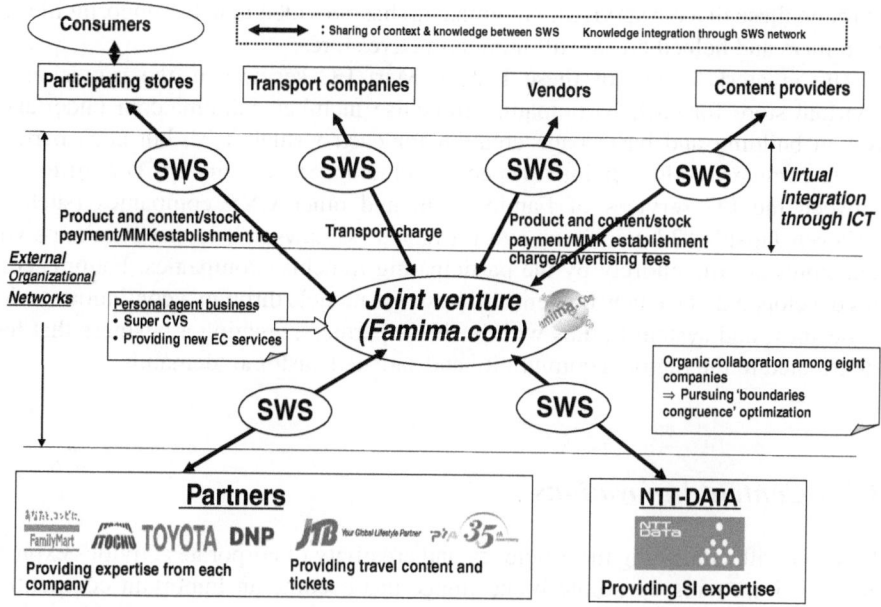

Fig. 3.5 Organizational networks centered on Famima.com

congruence" (congruence with the environment and within the corporate system) through assessment and coordination of new business models. Finally, they
established the new joint venture, Famima.com (see Fig. 3.5).

At Famima.com, the Family Mart HQ and participating stores united to promote the new twenty-first century "Super CVS" convenience store (CVS) format,
based on the world's first EC franchise system. Famima.com publicized its website as an Internet shopping service, and provided a product lineup that transcends
the limitations of existing convenience stores based on the two categories of home
delivery-based lifestyle solutions and shop counter-based lifestyle proposals.

Famima.com also built a system to support electronic commerce and established
a shopping website. The electronic commerce system includes services using multimedia information kiosks [MMKs] supporting satellite communications deployed
in Family Mart stores throughout Japan. MMKs are used to selling digital content
such as music distribution and digital printing as well as ticket and travel service sales, and provide all kinds of information services, including used-car data.
They have also expanded and strengthened lifestyle proposal and solution functions. Furthermore, in a different sales approach to Family Mart's OCOD (open
cash on delivery) services, Famima.com has developed an electronic-commerce
(EC) receipt agency service, enabling product receipt and cash-on-delivery for
EC company products (where customer demands are high) at the store counter
using a Family Mart distribution system. Moreover, the company has developed
a joint store-based ATM banking service in cooperation with Internet companies.

Through these diverse services, Famima.com has provided new EC opportunities to customers and helped create still more attractive stores.

The greatest feature of these Family Mart EC services is the provision of a virtual store for each participating franchise member. Famima.com undertakes system building and basic page creation for each virtual store, but each participating company adds individual aspects, including product lineup. The difference between the EC services of Famima.com and other CVS companies (such as 7-Eleven Japan's "7dream.com" and Lawson's "@Lawson") is that Famima's virtual stores are run entirely by the participating franchise companies. Famima.com has developed diverse new content and business models through a continuous series of business and system tie-ups with business partners, creating a business that has contributed to the regional community and satisfied customer demands.

3.3.5 Content Innovations

Together with enhancing the efficiency and creativity of corporate activities exploiting ICT, improving corporate brand image has become an important competitive element. Thus it is essential for each company to carry out efficient and effective advertising activities based on brand strategy. Meanwhile, the environment surrounding the advertising world, including the changing media structure, increasingly sophisticated client demands, and the increasing complexity of the advertising business, is ushering in an era of major reform. Under these conditions, demands to achieve still greater value-added communication in the advertising businesses of the future have been added to the real-time and efficiency demands of today. Recognizing this situation, in July 2007 Hakuhodo Inc. and NTT-DATA agreed on comprehensive cooperation based on the e-ad platform concept. Then in December 2007 they established the Ad Platform Inc. joint venture as an umbrella company, with investment ratios of 60 percent from Hakuhodo and 40 percent from NTT-DATA.

Ad Platform drew out the core competences that are the strength of the value chains for each of Hakuhodo's and NTT-DATA's vertical boundaries, established the new business of the "e-Ad Platform" concept, and expanded the horizontal boundaries. The NTT-DATA platform and Hakuhodo content innovators formed SWS aimed at collaboration, and targeted optimization of "boundaries congruence" (congruence with the environment and within the corporate system) by assessing and coordinating new business models. Finally, they established Ad Platform, Inc. as a new joint venture (see Fig. 3.6).

This e-Ad Platform concept builds a common ICT-based industry platform related to common industry operations (such as progress and materials management) for each business task, including production, media, and accounting. It then develops new business models on that platform, and aims to provide open access to relevant companies. The movement to create other industry standards has promoted useful links, and companies in the ad industry have enabled still greater focus

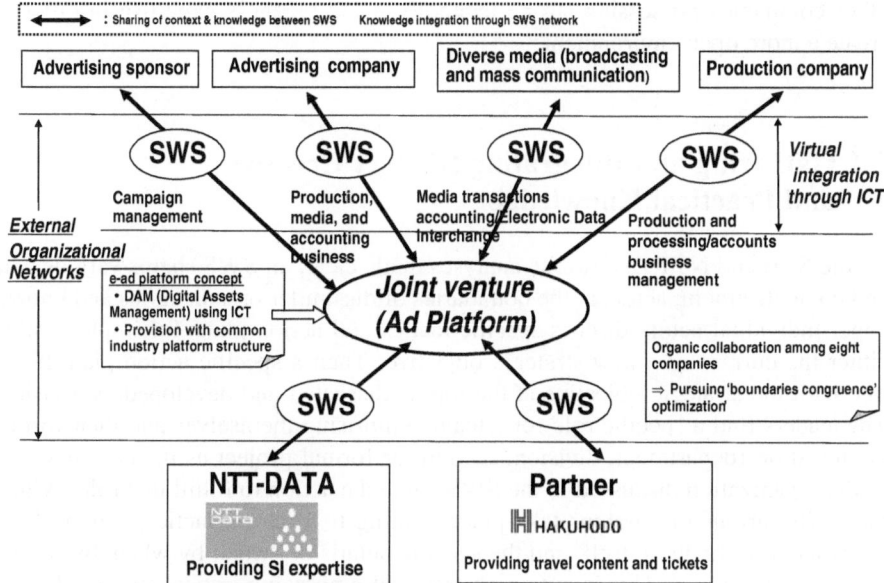

Fig. 3.6 Organizational networks centered on *Ad Platform*

on unique value-added through creative business and other means. This has been achieved by realizing the concept of promoting a common platform crossing sectors of the entire industry.

In specific terms, Ad-Platform builds a common industry hub centered on DAM (Digital Assets Management), undertakes production business including all kinds of media, graphics ads, CF, and poster catalogues, and assembles and reviews general digital workflow for business related to each of these categories. It then exploits innovation networks applying ICT to provide cheaper, higher quality advertising services through vertical integration to industry-related players, centered on the advertising companies.

At e-Ad Platform, ad companies, the media, and advertising sponsors share all data necessary for advertising production. This helps boost efficiency in the creation of advertising. Based on the shared data, workflow systems are created around various indicators and procedures related to the advertising business, such as sponsor orders, media requests for confirmation, and editing commissions. This enables the average 40-day period required for ad production to be reduced by two-thirds.

As a company providing DAM business, Ad-Platform consolidates all kinds of digital assets including content, media data, and marketing data, and provides efficient workflows integrating complex media and production work. Liberated from the complicated operations involved in the former manual procedures, ad companies using Ad-Platform can concentrate on planning while enhancing services for the sponsor and reducing delivery times and costs. This structure is effective for companies in related industries. Ad-Platform aims to cooperate extensively with other companies possessing strong technology and business expertise in order to create a more open environment.

3.4 Gathering and Integrating Distinct Creative and Practical Knowledge

As the SCE and NTT-DATA case analyses made clear, an SWS shares various data and contexts among actors at the boundaries of dissimilar organizations and knowledge. Individual actors discuss specific tactics and action plans for implementing either the current or a new strategic objective. Then a specific action plan, like a tree with the strategic objective at the top, is dissected and developed as a vision. The leaders find a specific role for a team comprising themselves and their formal organization (department, division, section, or formal project as the case may be) as the organization discussed in the SWS, and take action to fulfill that role. Within the SWS various discussions take place relating to strategy, tactics, action plans, the content of the item itself, and the specific details of "what, by when, by whom, to whom, and how." This is not an abstract, all-embracing critical strategy for the actors, but it becomes important that "strategy as practice" be implemented specifically (e.g., Pettigrew 2003; Whittington 2004). Within the SWS, the actors decide on key issues through constructive dialog and "creative collaboration." Then the issues that have been decided in the SWS are specifically tackled by the actors in the formal organizations to which they belong.

With special attention to the questions "how" and "by whom" a task will be done, actors must endeavour to form new SWS as they dynamically intensify the collaboration of the actors inside and outside the company and the existing SWS. A point of particular importance in forming a new SWS is the need to transmit the context and knowledge of other SWS with which one is associated to the prospective members of the new SWS. Transmission alone is not enough, however; new meaning based on the context to be communicated must also be produced (Nonaka and Takeuchi 1995), and specific pragmatic issues tackled (or executed by partners) in collaboration with new actors on the basis of this newly produced meaning. SWS are formed out of the dynamic changes to context that take place in this way. This produces new contexts and meaning, and further new SWS are formed. The formation of new SWSs produces new issues and action plans (or changes the action plan already being executed), and the actors implement the plan allotted for the team belonging to their formal organization.

The actors then appropriately confirm the extent to which the action plan has been attained within the SWS and discover still more problems and issues. Aiming to execute a problem-solving action plan, they reinforce execution in still more formal organizations and deliberately form new SWS as required. The actors participate in several SWS, and arrange for context and knowledge to be shared among the SWS. Middle management, especially, needs to commit to multiple SWS to benefit from knowledge integration by means of SWS networking. These SWS and networked

SWS exist in one of the formations of invisible human networks operating in the background of corporate and organizational network structures. At various times in various spaces (physically in the office or in cyberspace) actors share contexts with SWS and networked SWS and generate new meaning. These form the basis for strategies and action plans, which are then transferred to specific action through the formal organization.

The formation of such SWS and networked SWS (the practitioners' network concepts) becomes the trigger that creates knowledge in the form of new innovations and solutions to new problems and issues. The formation of SWS and networked SWS promotes enhanced recognition capabilities for new change among practitioners. The triggering of practitioners' new ideas on environmental change adaptation is unlikely to arise from the information processing model of daily, routine activities (e.g., Carlile 2004). SWS and networked SWS comprising members from different backgrounds and specializations, however, gather and integrate individual practitioners' creative and practical knowledge, and become the trigger for creating the practice processes of improvisational, emergent, and deliberate thoughts and action. Then practitioners faced with new environmental changes ask themselves how they can actively bring about a new process of change in themselves, their companies, and their organizations, and how they can create new business models. Practitioners in each organizational formation (formal organizations, SWS, and networked SWS) link these questions to formulating and implementing specific action plans (both emergent and deliberate).

This gathering and integration of distinct creative and practical knowledge simultaneously enhances the originality of the former and the efficiency of the latter. "Creative knowledge" refers to the ability of individual practitioners in different specialist fields to create original, path-breakthrough ideas, and becomes a source of new breakthrough (radical and discontinuous) innovation. The originality of creative knowledge becomes the engine creating new business models (including products and services) that either adapt to new, large-scale environmental change or actively create such changes. Thus creative knowledge becomes the ideal foundation on which practitioners conceive, propose, and formulate environment adaptive and creation strategies.

"Practical knowledge" has at its foundation path-dependent, tacit knowledge rooted in people's past experience, and the continuous accumulation of good-quality practical knowledge becomes the source of incremental innovation.[1] The efficiency of practical knowledge enhances the speed and degree of completeness that is ideal for specifically realizing new business models proposed and formulated on the basis of practitioners' high-quality experiential knowledge. Thus "practical knowledge" becomes the foundation for practitioners specifically achieving environment adaptive and creation strategies.

---

[1] "Practical knowledge" also includes a sense of "practical wisdom," and comprises "the capability to display the behavior and performance best suited to achieving visions aimed at truth, virtue, and beauty, and taking decisions and acting in order to realize overall optimization within specific scenarios" (Kodama 2007b).

As Chaps. 6 and 7 mention in detail, the evolution of this creative and practical knowledge enhances practitioners' ability to recognize change. Thus the gathering and integration of distinct creative and practical knowledge enhances the "recognition capability" of practitioners recognizing changes in the management elements of strategy, organization, technology, operation, and leadership within a corporate system possessing distinct contexts, and recognizing disparities among these elements. Through this advanced recognition capability, practitioners become aware of specific processes of changes adapting to environmental change and create strategy, organization, technology, operation, and leadership to achieve congruence among these individual management elements (mentioned below).

3.5 Optimizing Boundaries Congruence for Business Architecture

Next I will consider the business architecture of SCE and NTT-DATA. As mentioned in Chap. 2, these two companies managed to achieve congruence of the environment and corporate system and congruence of the individual management elements within the corporate system.

3.5.1 Optimizing Boundaries Congruence in SCE

SCE set out new environment creation strategies by gathering and integrating distinct creative and practical knowledge in the marketing and technology areas of the games console market, and beat out competitors to build a new value chain for game software distribution exploiting 3CG and CD-ROMs. This vertical integration value chain comprises end users, software manufacturers, SCE, component manufacturers, and stores. SCE optimized vertical boundaries as corporate boundaries, and dynamically achieved congruence with the changing environment. Later, SCE set out an environment adaptive strategy (see Fig. 3.7), demonstrating creative and practical knowledge accumulated by cultivating new markets so as to adapt to changes in the competitive environment and guarantee market share.

To achieve congruence among management elements within the corporate system, SCE simultaneously implemented distinctive technology, software, and distribution and sales strategies adapted to environment creation strategy, invested in new models (PlayStation-2, PSX, PSP, and PlayStation-3) as an environment adaptive strategy aimed at adapting to the competitive environment, and promoted a supporting software lineup strategy. The business models pursued by SCE are always cutting-edge, and are constantly evolving the technologies (in the broad sense of the word) of new supply chains in order to achieve games software that maximizes the capabilities of its hi-tech consoles and those of its creators while achieving business success. SCE's distinctive approach to reliably implementing

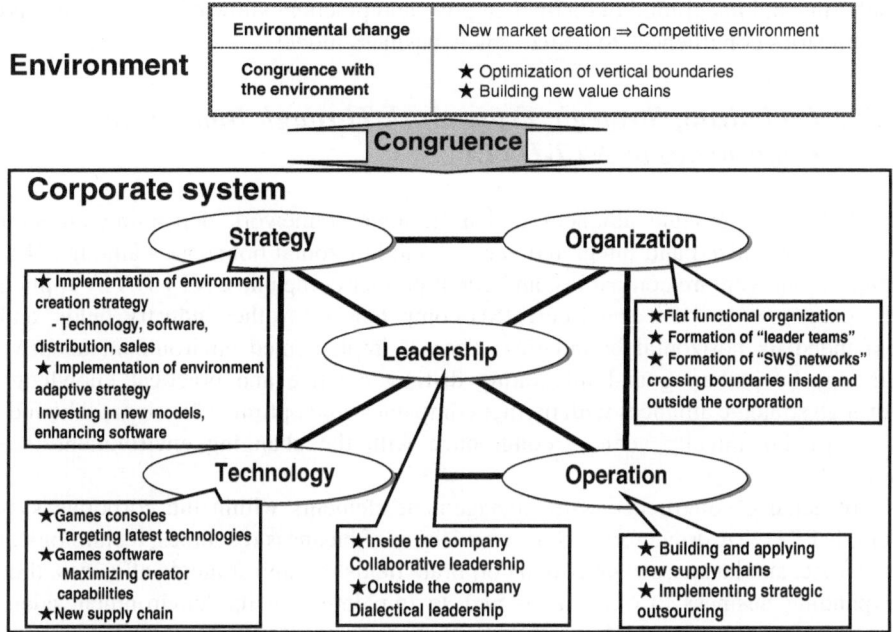

Fig. 3.7 SCE's congruence model

these strategies and technologies involves realizing games consoles exploiting superior knowledge from around the world by means of joint development (with Toshiba for the PS2, and with Toshiba and IBM for the PS3) through strategic cooperation; building ground-breaking supply chains ideal for distributing these products; and reliably implementing operational series from product planning, development, production, and distribution through sales.

Features of SCE organizational formation enabling the implementation of these strategies, technologies, and operations involve creating "leader teams" as SWS comprising specialized groups based on flattened, function-classified organizations while also creating integrated networks of SWS accumulating diverse knowledge distributed outside the company. This dynamic organizational formation creating SWS networks integrates knowledge distributed inside and outside the company, and becomes a platform for realizing new business models.

The task of implementing congruence among the individual elements of these strategies, organizations, technologies, and operations falls to the leadership of innovative practitioners, centered on SCE. These practitioners demonstrate "collaborative leadership" focused on leader teams within the company, and display mutual capability synergies. Outside the company, meanwhile, it is important for SCE to display not only vision but also "dialectical leadership" in forming win-win related structures among stakeholders for creating a new games market through PlayStation and building vertically integrated, SCE-centered value chains. In SCE's case, the formation of company-crossing SWS and networked SWS promoted the gathering and integration of distinct creative and practical knowledge, and achieved "boundaries congruence" to realize a new business model.

3.5.2 *Optimizing Strategic Framework Through Boundaries Congruence in NTT-DATA*

NTT-DATA's strategic alliances and a strategic framework depending on joint ventures optimized and integrated vertical and horizontal boundaries among NTT-DATA, joint venture companies, and client partner companies. NTT-DATA applied the core modules of its own core (SI) competences to other industry fields, and expanded its horizontal boundaries. Then it implemented environment adaptive strategy by gathering and integrating distinct creative and practical knowledge through strategic alliances with partner companies, and optimized horizontal boundaries to dynamically achieve congruence with the changing environment (see Fig. 3.8).

To achieve congruence with management elements within the corporate system, NTT-DATA drove forward the acquisition of business resources, technological expertise, and business expertise in corporate fields through strategic alliances, thus expanding business to other areas and so implementing its "environment adaptive strategy." It also reallocated resources, applying its own SI competences to other fields. Collaboration with the joint-venture parent companies (NTT-DATA

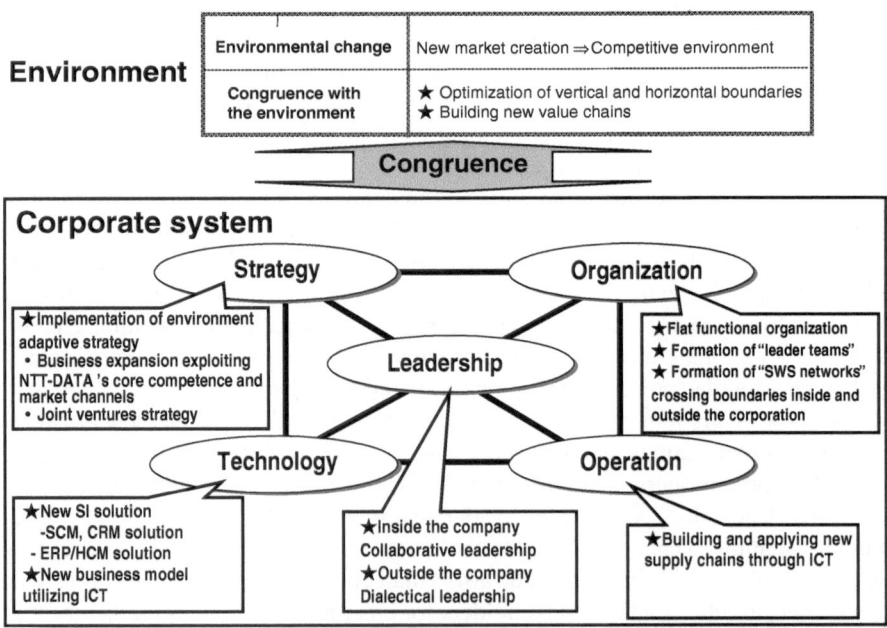

Fig. 3.8 NTT-DATA's congruence model

and investment companies partnered with NTT-DATA) also enabled optimal allocation and application of resources for the individual investment companies' vertical boundaries. Then the building of new value chains for the joint venture companies and investment companies' vertical boundaries created new awareness and motivation among practitioners, resulting in new ICT business and strengthening the corporate competitiveness of individual stakeholders. NTT-DATA's establishment of joint ventures from strategic alliances with partner companies promoted corporate boundaries congruence among NTT-DATA, joint venture companies, and client partner companies while enhancing the operational capability to implement environment adaptive strategy. The joint venture companies and NTT-DATA then collaborated with stakeholders to build value chains for new business model structures and ground-breaking supply chains through virtual integration exploiting value chain and ICT, and reliably implemented a series of operations from product planning, development, production, distribution, and sales through to after-service.

Features of NTT-DATA organizational forms that enabled implementation of these strategies, technologies, and operations include the formation of leader teams as SWS comprising division-crossing models based on flattened, functionally classified organizations together with the comprehensive networking of SWS nodules comprising diverse knowledge distributed outside the company in the form of partner companies and joint ventures. The dynamic organizational formation of these "networked SWS" merged knowledge distributed inside and outside the company (gathering and integrating distinct creative and practical knowledge), and became a platform for realizing new business models.

Congruence among the individual management elements of these strategies, organizations, technologies, and operations was then implemented by the leadership of innovative practitioners centered on NTT-DATA, joint venture companies, and partner companies. These practitioners promoted the gathering and integration of distinct creative and practical knowledge through the formation of SWS and networked SWS inside and outside the company, and displayed "dialectical leadership" aimed at co-creating business models based on collaborative leadership and new values, and building win-win-win relationships.

References

Albrinck, J., Horney, J., Kletter, D., Neilson, G. (2001). Adventures in corporate venturing. *Strategy and Business*, 22, 119–129.

Block, Z. (1982). Can corporate venturing succeed?. *Journal of Business Strategy*, 3(2), 21–34.

Block, Z., MacMillan, I. C. (1993). *Corporate Venturing: Creating New Business Within the Firm*. Cambridge, MA: Harvard Business School Press.

Bruch, H., Ghoshal, S. (2004). *A Bias for Action*. Boston, MA: Harvard Business School Press.

Bryson, J., Crosby, B. C. (1992). *Leadership for the Common Good: Tackling Public Problems in a Shared-Power World*. San Francisco, CA: Jossey-Bass.

Burgelman, R. A. (1983). A process model of internal corporate venturing in the diversified major firm. *Administrative Science Quarterly*, 28, 223–224.

Burgelman, R. A., Välikangas, L. (2004). Managing internal corporate venturing cycles. *Sloan Management Review*, 46(4), 26–34.

Carlile, P. (2004). Transferring, translating, and transforming: an integrative framework for managing knowledge across boundaries. *Organization Science*, 15(5), 555–568.

Chrislip, D., Larson, C. (1994). *Collaborating Leadership: How Citizens and Civic Leaders Can Make a Difference*. San Francisco, CA: Jossey-Bass.

Cramton, C. (2001). The mutual knowledge problem. *Organization Science*, 12, 346–371.

Gompers, P. A., Lerner, J. (1999). *The Venture Capital Cycle*. Cambridge, MA: MIT Press.

Kodama, M. (1999). Strategic innovation at large companies through strategic community management-an NTT multimedia revolution case study. *European Journal of Innovation Management*, 2(3), 95–108.

Kodama, M. (2001). Creating new business through strategic community management. *International Journal of Human Resource Management*, 11(6), 1062–1084.

Kodama, M. (2002). Transforming the old economy company to new economy. *Long Rang Planning*, 35(4), 349–365.

Kodama, M. (2004). Strategic community-based theory of firms- case study of dialectical management of NTT DoCoMo. *Systems Research and Behavioral Science*, 21(6), 603–634.

Kodama, M. (2007b). *Knowledge Innovation – Strategic Management as Practice*. Cheltenham: Edward Elgar Publishing.

Kodama, M. (2007c). *Project-Based Organization in the Knowledge-Based Society*. London: Imperial College Press.

Leonard-Barton, D. (1995). *Wellsprings of Knowledge: Building and Sustaining the Sources of Innovation*. Boston, MA: Harvard Business School Press.

Nonaka, I., Takeuchi, H. (1995). *The Knowledge-Creating Company*. New York: Oxford University Press.

Pettigrew, A. (2003). Strategy as process, power and change. in *Images of Strategy*. Cummings, S and Wilson, D. (eds.). London: Blackwell.

Shapiro, C., Varian, H. R. (1998). *Information Rules*. Boston, MA: Harvard Business School Press.

Star, S. L. (1989). *The structure of ill-structured solutions: Boundary objects and heterogeneous distributed problem solving.. Resding in Distributed Artificial Intelligence*. M. Huhns, L. Gasser, (eds.). Menlo Park, CA: Morgan Kaufman.

Vangen, S., Huxham, C. (2003). Nurturing collaborative relations, building trust in interorganizational collaboration. *Journal of Applied Behavioral Science*, 39(1), 5–31.

Von Hippel, E. (1997). Successful and failing internal corporate ventures: an empirical analysis. *Industrial Marketing Management*, 6, 163–174.

Whittington, R. (2004). Strategy after modernism: recovering practice. *European Management Review*, 1(1), 62–68.

Chapter 4
Developing New Broadband Services by Dynamic Collaboration Through Strategic Boundary Networks: A Case Study of NTT DoCoMo

4.1 Innovation Through Synthesis of Exploration and Exploitation

This chapter looks at a case of marketing innovation in Japan's fast-developing mobile phone business. The chapter has two key points of focus. It examines how the presence of dual networks (exploratory and exploitative) for major communications carrier NTT DoCoMo ("DoCoMo" from here on) helped to achieve a synthesis of environment creation and environment adaptive strategies and greatly expanded the mobile phone business markets. It also investigates the evolutionary process of DoCoMo's dynamic strategic management, and analyzes the dynamism of the congruence between the environment and corporate system and among individual management elements within the corporate system achieved by DoCoMo at each innovative phase.

For the kind of company mentioned in Chap. 2 to realize corporate innovation streams and maintain a constant competitive edge in the market, it must simultaneously establish practices to grow existing business and address the theme of future business development aimed at the acquisition of new strategic positions (e.g. Markides 1999; O'Reilly and Tushman 2004). Organizations aiming to grow their existing business (perhaps by improving existing products and services, and expanding profits through improvements and upgrades) must formulate and implement business planning in a short time frame. Actors in organizations of existing business activities can learn to predict business change and modify their practice to a certain degree. Actors with a base of tacit knowledge accumulated from such sources as path-dependent technology and sales are promoting exploratory practice as daily routines (Nelson and Winter 1982; March 1991).

Exploitation refers to the routine behavior involved in refining current organizational capabilities and improving the performance of current organizational routines. With exploitative practice, daily organizational learning aimed at incremental innovation (Nelson and Winter 1982; Tushman and Anderson 1986) is required of the organizational actors.

In their future exploratory practice (March 1991), companies must go beyond new scientific research and technical development to construct new business models

M. Kodama, *Boundary Management*, DOI 10.1007/978-3-642-03789-4_4,
© Springer-Verlag Berlin Heidelberg 2010

through marketing. Actors must boldly face market uncertainty and risk, and seek out new business by means of experimentation, incubation, and trial and error.

Actors promoting exploratory practice must emphasize architectural innovation from technological knowledge to effect radical change in technological design (Henderson and Clark 1990) and radical innovation from scientific knowledge comprising the development of core technologies based on new theories and principles (Fleming and Sorenson 2004). But it is also important for actors and organizations in practical businesses to emphasize new knowledge from a marketing focus that looks beyond technological and scientific knowledge to radically transform existing business systems. Put in another way, new marketing knowledge is needed to dramatically transform the existing value network (Christensen 1997) and value chain (Porter 1985), and to create new business models. Marketing knowledge that brings the customer close has the important capability of creating radical innovations leading, for instance, to new business creation or lifestyle changes. I would like to use the term "marketing innovation" for innovation driven by marketing knowledge (the mobile phone business case in this chapter is an example of marketing innovation that brings the customer close).

The thinking of organizational leaders aiming to acquire marketing knowledge with a customer-oriented stance must incorporate the following: freeing themselves from existing mental knowledge models (see Spender 1990; Banker 1993); a temporal and spatial system where diverse, heterogeneous knowledge intersects (Johansson 2004); new concepts to leverage and stretch existing resources and capabilities from strategic intent (Hamel and Prahalad 1989); avoidance of competency traps (Levitt and March 1988; Martines and Kambil 1999) and core rigidities (Leonard-Barton 1992, 1995); destruction of the organization's information filter (Henderson and Clark 1990); destruction of the organization's power structure and "creative destruction" of vested rights and resources; breaking down the organization's structural inertia (Hanna and Freeman 1984); breaking down successful personal experience and promoting strategic policies for future aims (Ackoff 1981); and building a hierarchy of imaginative capability (Hamel 1996; Mintzberg et al. 1998). Individual actors, moreover, must deliberately shake up their "thought worlds" (Dougherty 1992) based on individual backgrounds and specialization from the viewpoint of the customer. Organizational leaders also need the element of "disciplined imagination" (Weick 1989; Kodama 2003) to select the many exploratory strategies that have been discovered and make decisions appropriately.

With the implementation of the "environment creation strategy" mentioned in Chap. 2, "exploratory" strategy processes become especially important. Exploratory practice becomes the engine driving radical innovation to create new markets. With the implementation of "environment adaptive strategy," both exploitative practice based on existing abilities and exploratory practice adapting to new environmental change become necessary. The synthesis of these two practices creates incremental and radical innovation adapting to environmental change.

Observations from the case studies in this chapter have two key areas of focus. One is the dynamic process whereby DoCoMo introduced dedicated organizations based on projects separated from the existing bureaucratic, functional organizations, and formed industry-crossing "exploratory networks" achieving radical innovation through two strategies: an "environment creation strategy" for future business creation and an "environment adaptive strategy" adapted to major environmental change. This area also involves the formation of exploitative networks as constant routine activity, which is aimed at incremental innovation supporting environment adaptive strategies by bureaucratic functional organizations. The purpose of this innovation is to diffuse, improve, and define new business arising from the formation of the exploratory networks. The exploitative networks emphasize the aspects of expanding and growing existing business, while the exploratory networks emphasize the creation of new business.

The formation of these organically consolidated "dual networks" comprising exploratory and exploitative networks, and the practitioners' practice process of formulating and implementing environment creation and environment adaptive strategies is greatly enlarging the business market for mobile phones through dialectical recursive interplay (Giddens 1984; Barley 1986; Barley and Tolbert 1997) between the environment and markets (structure) on the one hand and organizations and individuals (practice) on the other. Corresponding to the nodules of these dual networks are the presence of "small-world structures" (SWS) (also called "small-world networks" [SWN]) comprising the leaders and managers of project-based and functional organizations crossing organizational boundaries within DoCoMo. The SWS that take on the role of dual network formation form "leader teams" (LT) from managers within the organization, and play a central role in dialectically implementing environment creation and adaptive strategies.

The second area of focus is the evolutionary process of DoCoMo's dynamic strategic management. In each innovational phase DoCoMo demonstrated the dynamism to achieve congruence of the environment and corporate system and of the individual management elements within the corporate system.

4.2 Case Study: Mobile Phone Business Innovation

This case involves DoCoMo's i-mode innovation and its third-generation FOMA mobile phone service, which DoCoMo aimed to develop and distribute as a global pioneer of Internet and multimedia related mobile communications services.

What is notable about this case is that while DoCoMo was restricted by the limitations of the market environment, dual networks (exploratory and exploitative) holding diverse knowledge formed spontaneously within and outside the organization. Here I will analyze the dynamism involved in building a new market environment integrating dual networks by creating and distributing new services in Japan and elsewhere while developing new services for the next-generation system.

4.2.1 DoCoMo's Innovations

The main driver behind the expansion of mobile computing's potential and usability is data communication from mobile phones. Today's mobile phones have progressed from "portable handsets" to "information terminals" thanks to Internet-accessing technology represented by DoCoMo's i-mode. In the mobile Internet field, Japan has a lead of two to three years over the US and Europe. US journalists, moreover, suggest that Japan's wireless Internet access system, which is wildly popular in Japan, has strong global potential.

i-mode currently serves 40 million subscribers in Japan, while the i-mode business model is spreading worldwide. Japan has also launched the fast i-mode service through the pioneering FOMA third-generation mobile phone service centered on DoCoMo, and the new system is being further developed for the global market. I will describe the three phases of DoCoMo's innovation process chronologically (see Fig. 4.1).

4.2.2 Phase 1 (1992–1998): The Challenge of Voice Communication

In 1992, the mobile communications business of NTT (Nippon Telegraph and Telephone Corporation), Japan's largest telecommunications enterprise, was spun off to ensure a fair, competitive telecommunications market (See 4.1). That was the birth of NTT DoCoMo. The government's objective was for DoCoMo to split from NTT, and to launch into the competitive market with its parent company barred from support of any kind. The first fiscal year returned a deficit, however, as net sales fell below the previous year's figure due to poor mobile phone sales.

DoCoMo was launched as a small- or medium-sized business. It encouraged initiatives by creating a climate where individuals always transcended organizational boundaries to share information and knowledge, and which promoted business oriented toward a shared sense of values and targets. The employees crossed these boundaries in individual communities of practice such as sales, development, technology, maintenance, and planning, and in-house communities developed to deal with urgent topics, consisting of informal projects, task teams, or the company as a whole.

Aiming to avoid negative cycles, DoCoMo's president Koji Oboshi's first task as CEO was for leaders and managers to create informal, cross-functional project teams from each division, including marketing, sales, technology, and equipment, that transcended organizational boundaries. The leaders and managers of the development and technology divisions that supported the mobile phone platform also undertook joint development by forming trial-and-error "exploratory networks" with mobile handset manufacturers. The aim was to surpass Motorola by developing the world's lightest high-function phones. (Fig. 4.2 illustrates the organizational structure of the mobile phone's first phase).

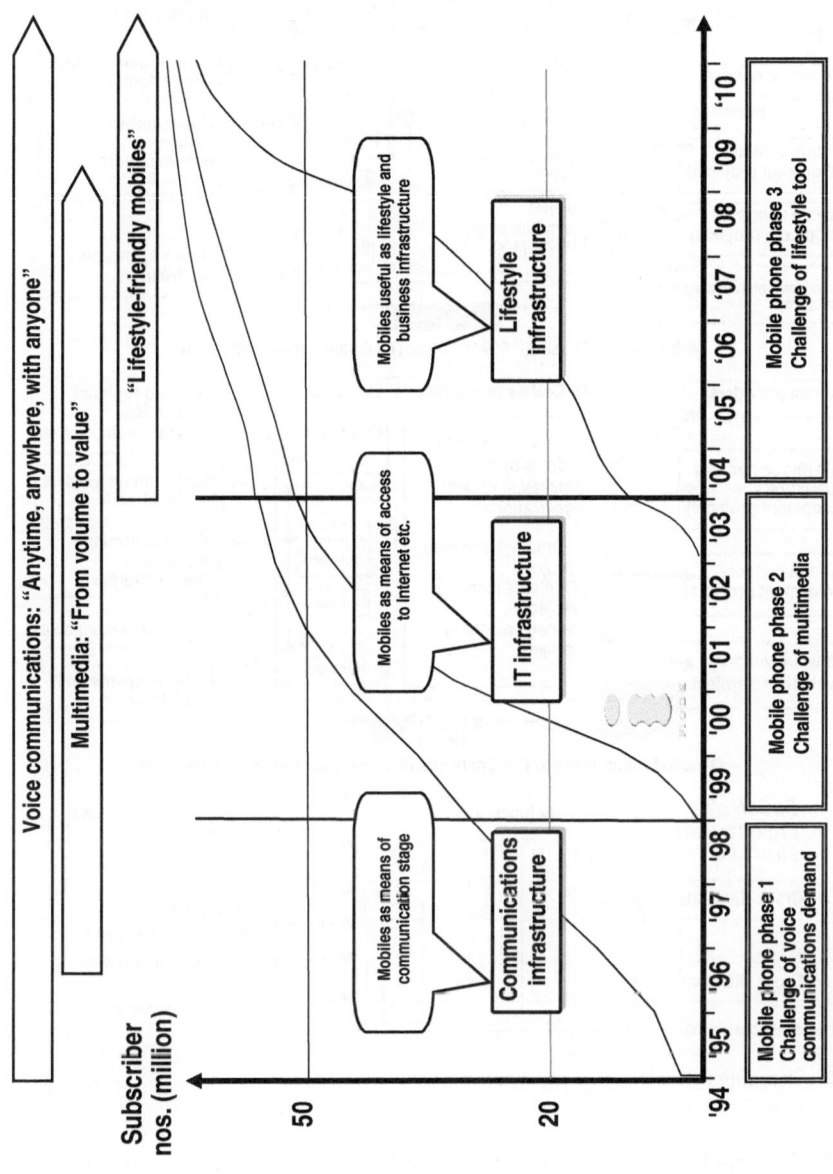

Fig. 4.1 TT DoCoMo inncvations

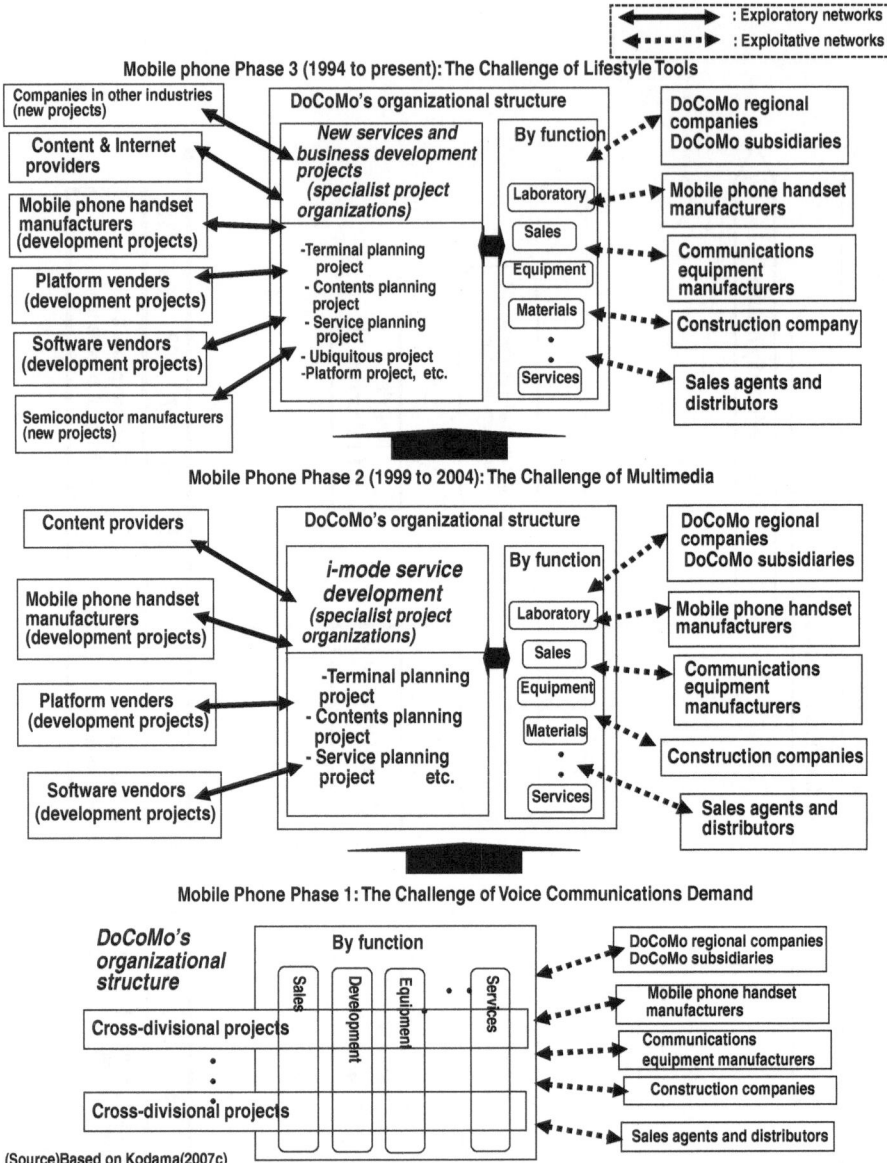

Fig. 4.2 Changing external organization networks

DoCoMo employees took swift action to address various themes. The expansion of the communications network, especially, created strategic 'exploitative networks' with communications equipment manufacturers as strategic partners, group companies, and affiliated companies (construction companies), and aimed to build equipment and develop networks that took into consideration a balance of timely demand and supply (see Fig. 4.2 for the organizational structure of the mobile phone's first phase).

Concrete measures were implemented to solve the issues one by one, with the result that sales quickly rose threefold, and the world's lightest mobile phone (developed by DoCoMo's technology group) took off. As the phone began to sell, economies of scale kicked in and the mobile phone grew cheaper. The mobile phone "rent back" problem arose, and Japan's Ministry of Posts and Telecommunications regulating authority sensibly demanded the adoption of an outright sale approach. Competition intensified as a result.

In addition, DoCoMo built its sales network by outsourcing to existing electrical goods outlets as agents and distributors (Fig. 4.2 shows the organizational structure at the mobile phone's first stage). This enabled DoCoMo to expand sales channels quickly, and so both recover market share and create new demand. Through this kind of strategic outsourcing and sales channel expansion, DoCoMo succeeded in growing the market from the conventional business to the personal use layer and creating new demand.

To recap, the DoCoMo startup at the mobile phone's initial phase found itself in an environment where restrictive conditions rendered it unable to sell. A new market (structure) for mobile phones arose through practice resulting from the formation of "exploratory networks" (cross-departmental projects) and "exploitative networks" (functional organizational networks) arising from a sense of crisis shared among the whole company. Meanwhile, newly created structures were driving competition, and led to new innovations and strategies in organizational networks, especially for NTT DoCoMo (see "Mobile Phone Phase 1," Fig. 4.3).

From the viewpoint of "corporate innovation streams" described in, Fig. 2.4, DoCoMo created new markets and competitive environments by implementing an "environment creation strategy" cultivating mobile phone markets. It also implemented a new "environment adaptive strategy" adapted to this competitive environment.

4.2.3 Phase 2 (1999–2004): The Challenge of Multimedia

Mr. Oboshi predicted as early as 1997 that the mobile phone subscriber growth curve for voice communications would soon reach saturation point (See 4.1). As a result, DoCoMo's profits would fall at some point, creating a sense of crisis that DoCoMo's growth would be endangered. This prompted Mr. Oboshi to turn his attention to the new data and image (multimedia) communications market, which was displacing that for voice communications (see Fig. 4.4).

In January 1997 Keiichi Enoki (formerly NTT DoCoMo's managing director and general manager of i-mode, currently CEO of NTT DoCoMo Engineering, Inc.), who worked as division head of NTT DoCoMo, was instructed by Oboshi to develop mobile multimedia services using mobile phones for the general user. Oboshi also instructed Enoki to gather talent by scouting outside the company and appointing

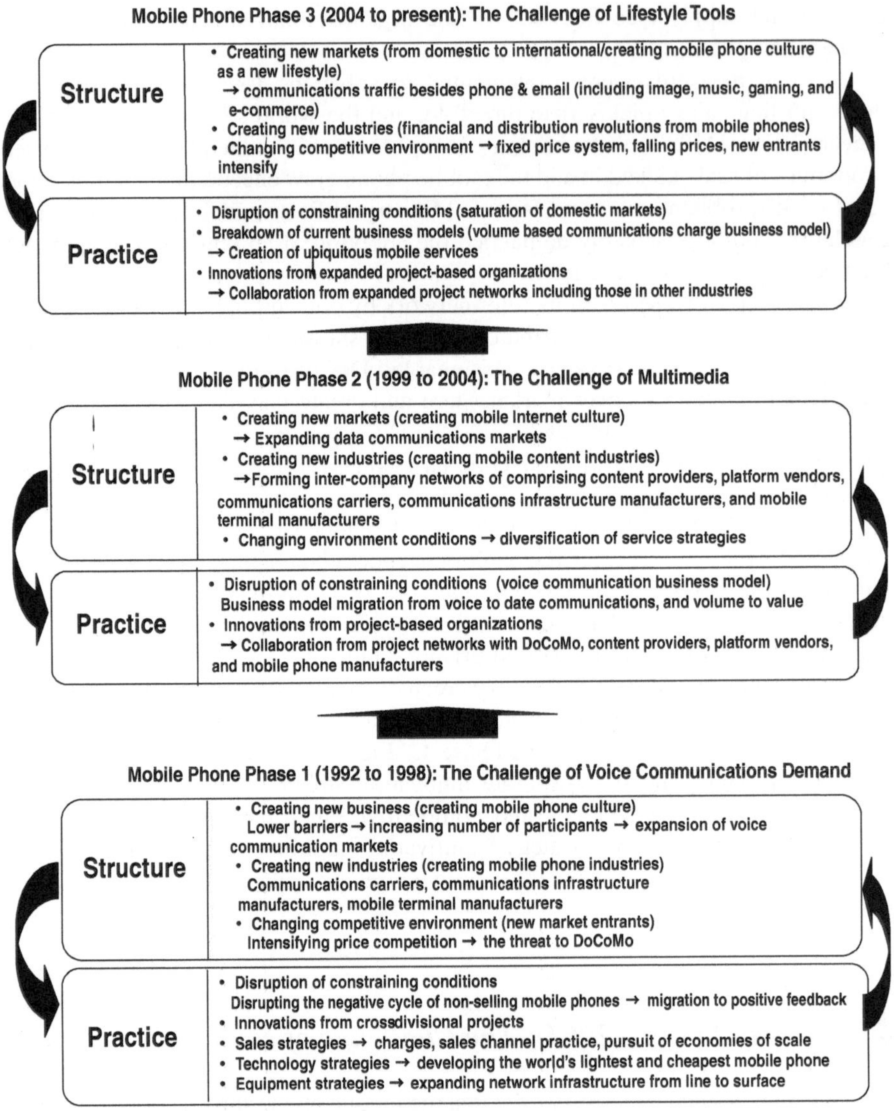

Mobile Phone Phase 3 (2004 to present): The Challenge of Lifestyle Tools

Structure	• Creating new markets (from domestic to international/creating mobile phone culture as a new lifestyle) → communications traffic besides phone & email (including image, music, gaming, and e-commerce) • Creating new industries (financial and distribution revolutions from mobile phones) • Changing competitive environment → fixed price system, falling prices, new entrants intensify
Practice	• Disruption of constraining conditions (saturation of domestic markets) • Breakdown of current business models (volume based communications charge business model) → Creation of ubiquitous mobile services • Innovations from expanded project-based organizations → Collaboration from expanded project networks including those in other industries

Mobile Phone Phase 2 (1999 to 2004): The Challenge of Multimedia

Structure	• Creating new markets (creating mobile Internet culture) → Expanding data communications markets • Creating new industries (creating mobile content industries) →Forming inter-company networks of comprising content providers, platform vendors, communications carriers, communications infrastructure manufacturers, and mobile terminal manufacturers • Changing environment conditions → diversification of service strategies
Practice	• Disruption of constraining conditions (voice communication business model) Business model migration from voice to date communications, and volume to value • Innovations from project-based organizations → Collaboration from project networks with DoCoMo, content providers, platform vendors, and mobile phone manufacturers

Mobile Phone Phase 1 (1992 to 1998): The Challenge of Voice Communications Demand

Structure	• Creating new business (creating mobile phone culture) Lower barriers → increasing number of participants → expansion of voice communication markets • Creating new industries (creating mobile phone industries) Communications carriers, communications infrastructure manufacturers, mobile terminal manufacturers • Changing competitive environment (new market entrants) Intensifying price competition → the threat to DoCoMo
Practice	• Disruption of constraining conditions Disrupting the negative cycle of non-selling mobile phones → migration to positive feedback • Innovations from crossdivisional projects • Sales strategies → charges, sales channel practice, pursuit of economies of scale • Technology strategies → developing the world's lightest and cheapest mobile phone • Equipment strategies → expanding network infrastructure from line to surface

(Source)Based on Kodama(2007c)

Fig. 4.3 Recursive interplay of structure and practice

from within DoCoMo, and to build a new organization. Oboshi delegated all power over personnel and resources to establish the new services.

Enoki gathered capable and remarkable personnel both from inside and outside DoCoMo (including the recruitment of content specialist Mari Matsunaga and Takeshi Natsuno from an IT venture company), and launched the project (in

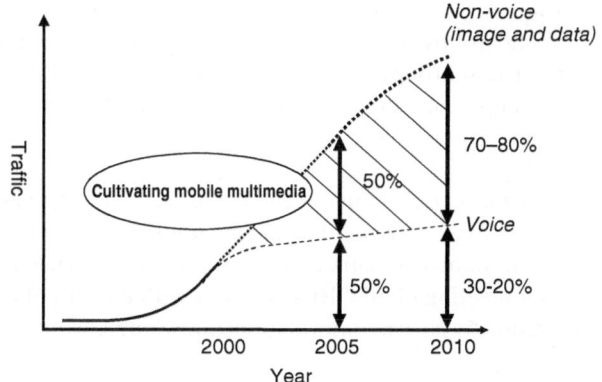

Fig. 4.4 Shift from voice to multimedia communications (image)

charge of Gateway) with a staff of around 10. By August 1997, a new organization, Gateway Business Division (GBD), had been launched with a staff of around 70. GBD undertook development of the new i-mode service as a project-based organization under Enoki's leadership. GBD employees, especially Oboshi and Enoki, shared visions and individual thoughts about the development of the new service.

4.2.3.1 Forming "Small-World Networks" (SWN) Within DoCoMo

Development of the i-mode mobile phone handset and server required cooperation among the development, technology, and equipment divisions of DoCoMo's functional organization. To begin with, however, opinion about the services within these organizations was negative, and disagreements over the services arose among personnel in the GBD and other divisions. Enoki stood at the front line of these contradictions and conflicts, and he resolved them constructively and productively by steering the divisions in the direction of tenacious dialectic debate and collaboration.

Enoki's leadership in gathering together all GBD members with the mission of making a success of the i-mode service, on which they had staked their professional pride, provided the driving energy behind the in-house organization.

Enoki displayed strong leadership aimed at realizing the i-mode service, acquired the understanding and agreement of the functional organization's leaders, and launched a liaison group to promote the introduction of an in-house mobile gateway service. This liaison group comprised all the leaders of the DoCoMo division, including the CEO and Enoki. It was positioned as a SWN to share information and knowledge at the top management level aimed at realizing i-mode services, and to offer dialog and the time and space for decision making aimed at business promotion.

Meanwhile, the project leaders, centered on Enoki, launched a total of seven working groups comprising middle management from the GBD and other organizations. These groups covered network servers, mobile phones, equipment construction, equipment maintenance, systems and sales, and content and application. These working groups were SWNs to sift and discuss specific issues and themes aimed at realizing the i-mode service. Every Tuesday, moreover, a task force specializing in i-mode handset development and i-mode servers (the Gateway Service Specifications Investigation Committee) would assemble to decide on service and technology specifications aimed at realizing the i-mode service. Tuesdays were also earmarked for a regular meeting of all GBD members with the aim of promoting discussion and collaboration by sharing information, knowledge, values, and feelings about the i-mode project.

Enoki and Natsuno seized on the following strategies aimed at a dramatic expansion of i-mode. First was a portal strategy for developing attractive new content using i-mode. Second was a terminal strategy for product development of new i-mode mobile phone terminals, including the addition of new functions. Third was a platform strategy to cultivate subscribers who use the handset as a platform as well as a mobile phone. These three business strategies were mutually connected and created considerable synergies.

Promoting this kind of business strategy requires the action of positively promoting "exploratory networks" (project networks) from strategic cooperation among various external partners and creating specific results. The organizational behavior that we should focus on here is the resonance of DoCoMo's new business strategy proposals with external partners and the establishing of projects with individual partner companies. DoCoMo deliberately builds its projects into exploratory networks with external partners. These exploratory networks host a continuous creative dialog oriented to the formation of new environments (structures) to establish and spread mobile Internet culture. Then various problem areas and issues are dialectically synthesized, and i-mode's new business concepts are created. A key feature of these "exploratory networks" in the second phase is their greater expansion outside the company.

4.2.3.2 Forming "Exploratory Networks" with CPs

Meanwhile, the GBD content planning project faced the major issue of how to acquire CPs with appealing content. The strategy adopted by content planning project leaders Matsunaga and Natsuno was to build win–win relationships leading to the success of both CPs and DoCoMo. When creating the i-mode content lineup, DoCoMo unilaterally bought content from specific CPs, but instead of taking a "tenant fee" from the CPs, both the CPs and DoCoMo acted from the opposed standpoints while sharing the concepts of risk and profits, creating the important element of establishing a win–win relationship. DoCoMo's first step was to have the CPs create first-class content and provide a platform to collect a content service usage fee as an agent, thus acquiring a service profit from the end user (see Fig. 4.5).

Fig. 4.5   i-mode data fee
collection agency system

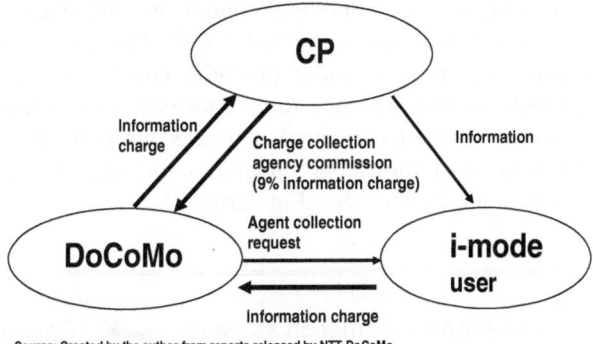

Source: Created by the author from reports released by NTT DoCoMo.

Matsunaga and Natsuno explained the concept of the win–win relationship to numerous CP representatives, fully explaining the situation and winning over the CPs to DoCoMo's way of thinking. DoCoMo and the CP values resonated, and "exploratory networks" aimed at establishing new business were formed. Through these networks, DoCoMo and the CPs jointly addressed the issue of developing content to make the end user truly happy while paying a content service fee. They discussed the speed, accuracy, and sustainability of content, and the viewpoint of end user satisfaction, and then went ahead to create appealing content that the end user would not tire of. Driven by Matsunaga and Natsuno, the GBD content planning project acquired a succession of effective CPs including mobile banking, credit card, airline, hotel, news, newspaper, and magazine providers. By the time of the i-mode service launch in February 1999, DoCoMo had acquired 67 CPs.

In this way, Enoki, Matsunaga, and Natsuno deliberately formed SWS of top and middle management within DoCoMo but centered on GBD, external exploratory networks with CPs, and exploratory networks for technology discussions with mobile phone handset manufacturers, thus building up positive and constructively creative dialogs aimed at forming the new environment (structure) for the i-mode mobile Internet market.

Within these exploratory and small world networks the gaps between individually held knowledge and the present environment (structure), as well as the conflict and contradictions among individuals, were dialectically synthesized, and the i-mode service concept was created. Aiming to launch this service, the in-house functional organizations of the GBD and DoCoMo cooperated effectively and moved the service concept toward completion at a higher strategic and analytical level, aided by production and consolidation of specific knowledge in such areas as i-mode compatible mobile handsets, i-mode server specifications, sales manuals, operation manuals, and content lineups. The construction of the i-mode promotion system and the detailed specifications of the i-mode service were also initiated in-house.

The distribution and establishment of new services, however, cannot be realized solely by exploratory networks (project networks) as exploration practices. Moreover, exploitative networks formed from in-house functional organizations (including sales, technology, equipment, and maintenance) as exploitation practices must function sufficiently well. DoCoMo firmly consolidates these dual exploratory (project) and exploitative (functional organization) networks in-house through small-world structures. DoCoMo also integrates the individually differentiated knowledge in dual networks through small-world networks. I call this SWN-inspired knowledge integration and organizational capability "network integrative competence" (See Fig. 4.6).

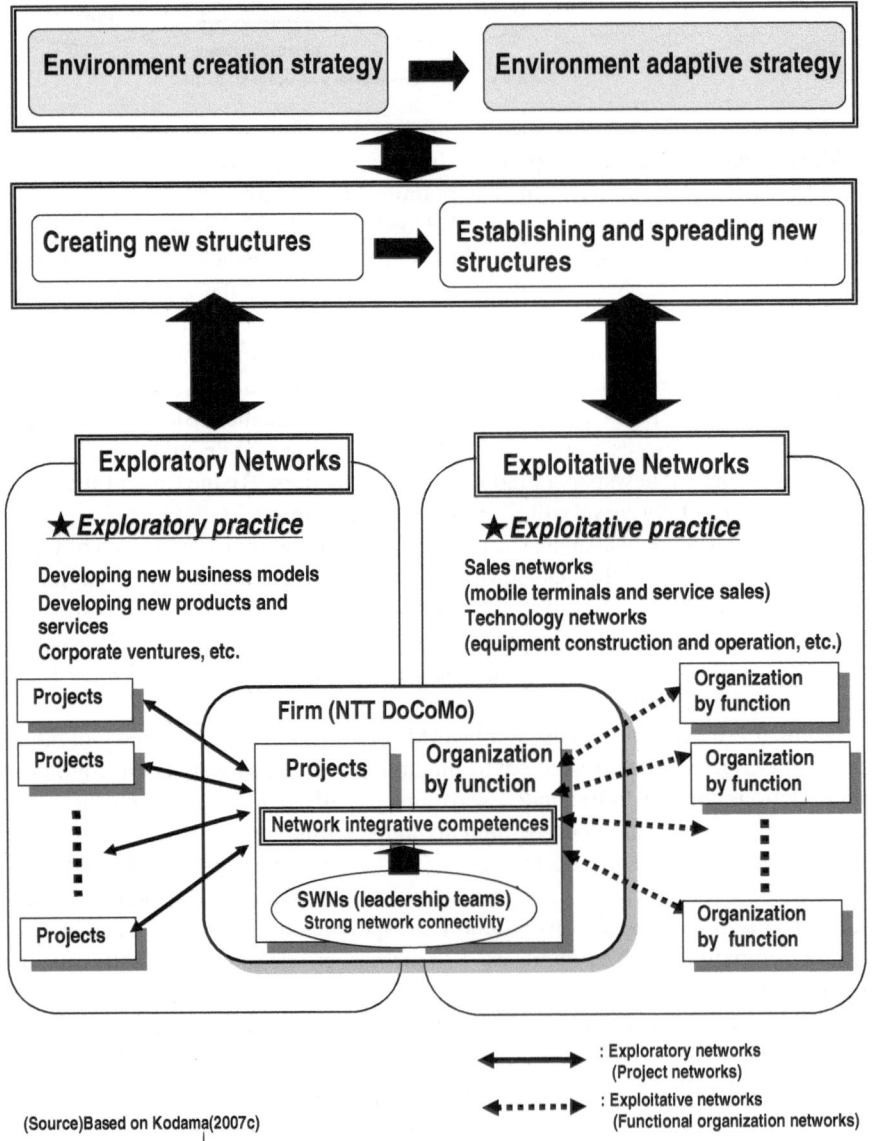

(Source)Based on Kodama(2007c)

Fig. 4.6 Recursive interplay of structure and networks

The i-mode service was launched in February 1999. By August 2000, i-mode subscriber numbers exceeded 1,000, and the world's first mobile phone Internet market was born. DoCoMo had created a new market, a new environment, and a new structure (see Mobile Phone Phase 2 in Fig. 4.3). Through i-mode, the mobile phone user could access online services, send and receive email, connect at will to the Internet as if on a PC, and obtain desired information "easily," "anywhere," "at any time." By realizing the concepts of "simplicity" and "requiring only a mobile phone handset," i-mode took the first step toward developing mobile multimedia.

The wildly popular i-mode system comprises four main system elements (see Fig. 4.7). The first is the i-mode compatible mobile phone handset. This is loaded with a web browser (using a subset of HTML) enabling users to enjoy high-speed access through the current third-generation FOMA system. The second is the packet communications network (the packets connect with the i-mode server). The third is DoCoMo's i-mode server network and web-connection functions. These execute content distribution, send, receive, and store e-mail, and user and handle CP administration. The fourth is content, provided by CPs to mobile phone users using the medium of the Internet and dedicated lines.

The rich content also contributed to the increase in i-mode users (see Fig. 4.8). In the background was the building of a win-win relationship with CPs. Acting as an agent for the content providers, DoCoMo constructed a content fee collection system that enabled CPs to collect small service fees from huge numbers of mobile phone users. The CPs and DoCoMo thus created a mutually profitable, win-win relationship whereby CPs provided attractive content for mobile phone users in an expanding virtuous cycle (see Fig. 4.9).

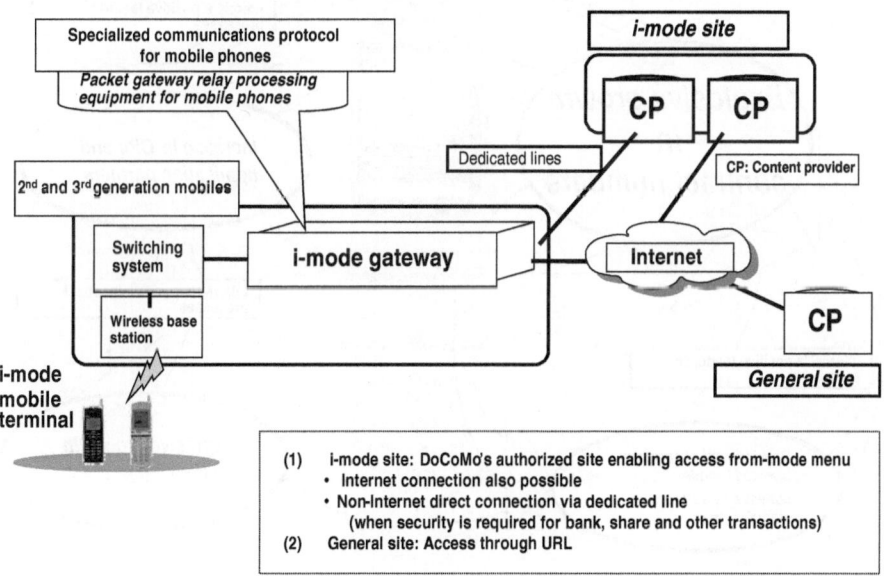

Fig. 4.7   i-mode service system structure and images

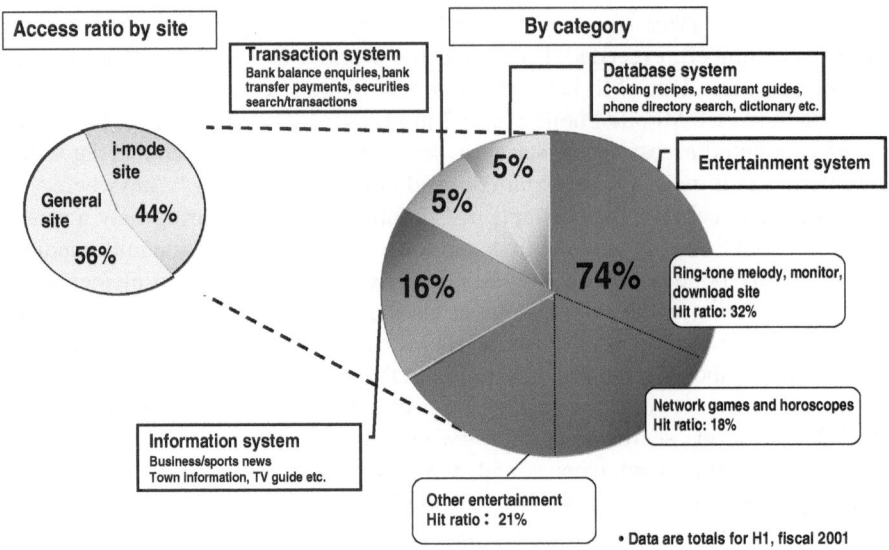

Source: Created by the author from reports released by NTT DoCoMo.

Fig. 4.8 Access ratios by i-mode site category

The i-mode portal offers a huge number of distinctive services including advertising distribution, sophisticated financing from Internet banks, development of new
java-enabled mobile handsets through the i-appli service (see Fig. 4.10), the i-area

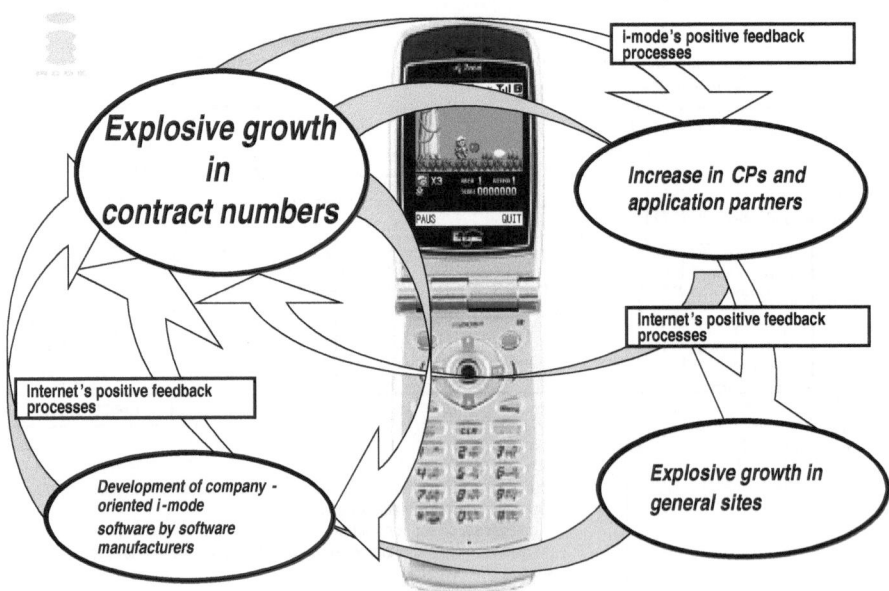

Fig. 4.9 i-mode's positive feedback

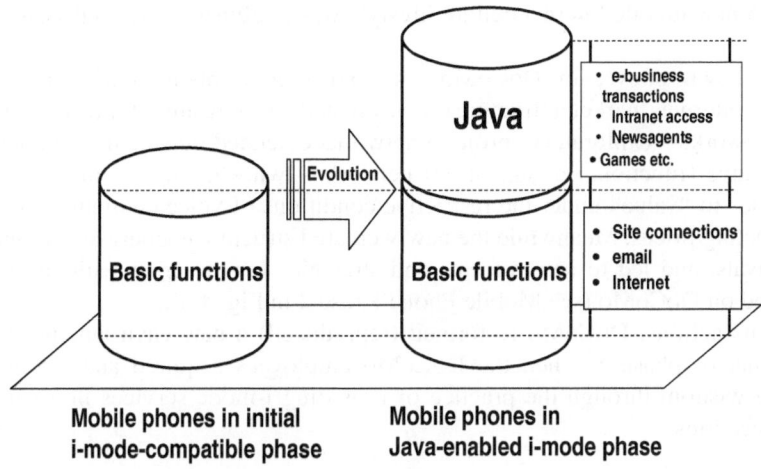

Fig. 4.10 Evolution of Java-loaded i-mode mobile phones

service, game machine links (such as PlayStation), the c-mode service (links to vending machines), convenience stores (Lawson), links to AOL email, and links to car navigation.

In December 2001, DoCoMo achieved the figure of 30 million subscriber contracts (see Fig. 4.11). i-mode was a great hit in Japan and had become a social talking

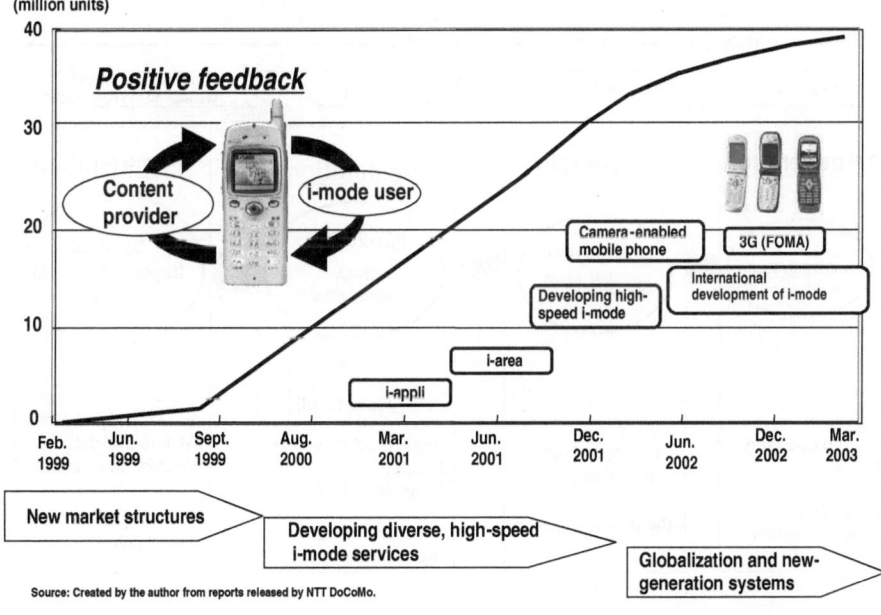

Fig. 4.11 i-mode innovation

point. A new mobile Internet culture lifestyle was established, especially among the young.

As shown above, in DoCoMo's second mobile phone phase the i-mode mobile Internet market (structure) was created as a result of practice through dual networks—exploratory (project) networks extended outside the company and exploitative (functional organization) networks—with the aim of migrating from "volume" to "value" under the restrictive conditions of voice communications and plummeting prices. Meanwhile the newly created structure encouraged competition with rivals, and led to new reforms and strategies for the organizational network centered on DoCoMo (see Mobile Phone Phase 2 in Fig. 4.3).

By its actions, DoCoMo re-formed (reproduced) a new environment differing from that of phase 1. Then the DoCoMo employees acquired and accumulated diverse wisdom through the practice of providing i-mode services in a variety of sales locations.

4.2.4 Phase 3 (2004 to Present): The Challenge of a Lifestyle Tool

KDDI, Vodafone (now SoftBank), and other DoCoMo rivals have also entered the mobile Internet service market and steadily made it more dynamic (See 4.1). The next challenges for DoCoMo, as a company that has achieved 40 million i-mode subscribers, were first, to expand i-mode distribution to foreign markets; second, to migrate from the then-current second-generation system to a third-generation system (see Fig. 4.12); and third, to realize the idea, close to DoCoMo's heart, of

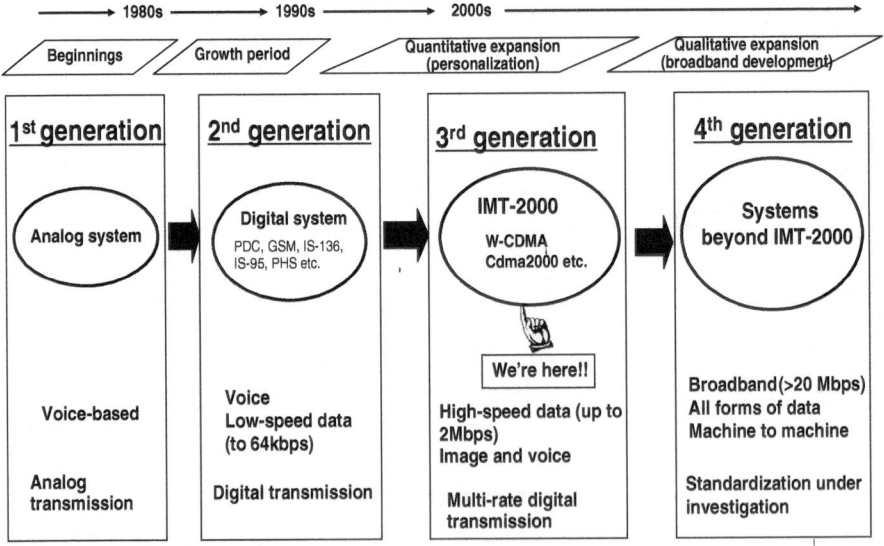

Fig. 4.12 Developing mobile communication systems

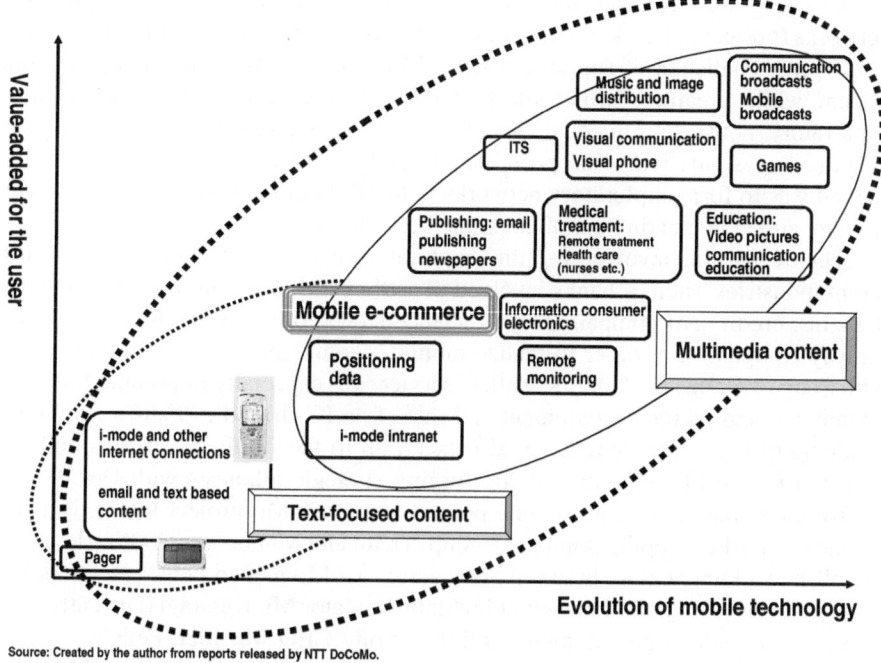

Source: Created by the author from reports released by NTT DoCoMo.

Fig. 4.13   Evolution of mobile content from connection to enjoyment and convenience

realizing a mobile phone "lifestyle tool" (See Fig. 4.13) represented by a mobile e-commerce service (the "mobile wallet" service).

Regarding the question of creating a new business model for i-mode's further development, Enoki and Natsuno always embraced the spirit of innovation without being complacent about the environment (structure) they had created. Furthermore, alongside the project-based organization (the Product and Service Division of i-mode's business headquarters, which later changed its name to the Product and Service Headquarters following reorganization) that Enoki and Natsuno belong to, the DoCoMo organization that took the credit for i-mode's success formed a vision of creating a new communication culture and realizing a new knowledge society through the twenty-first century mobile.

The knowledge (experience, skill, and expertise accumulated individually by Enoki, Natsuno, and the other employees) that led to i-mode's domestic success with a second-generation system arose from sharing among related divisions of each management level, including DoCoMo's functional organizations, and from new inquiry and speculation aimed at DoCoMo's further development.

Two major themes that dominated at DoCoMo during this phase were the growth of the domestic market resulting from the profitability of the current system and the interaction of new business risks from foreign markets and new-generation systems. A further theme was the question of how to establish a new mobile phone service as a lifestyle infrastructure.

DoCoMo's project-based and functional organizations formed global exploratory networks (project networks) both within DoCoMo and with external partners aimed at the great challenge of supporting DoCoMo's globalization and next-generation system while creating further lifestyle tools. The first challenge was to form exploratory networks aimed at the overseas development of i-mode and third-generation systems with European, Asian, and US communication carriers. The second was to form exploratory networks with different industries as a new service strategy aimed at creating a mobile phone lifestyle tool.

Specifically, this involved forming exploratory networks with companies in different industries, such as banks involved in settlements and commercial transaction domains, credit card companies, convenience stores, different kinds of stores, and railway companies, in order to realize mobile e-commerce services. Collaboration with Sony to realize the "mobile wallet" service was especially important. Sony had already worked on the development and sale of an IC chip (the "FeliCa chip"), but profitability was a problem. Sony also faced up to the challenge of risk, and displayed a forward-looking attitude by forging strategic alliances with DoCoMo to realize e-commerce through mobile phones. As DoCoMo project leader, Natsuno displayed skillful "tipping point leadership" (Kim and Mauborgne 2005). He finally established a joint venture between Sony and DoCoMo named FeliCa Network Inc. through top-level negotiations with PlayStation creator Mr. Kutaragi (currently CEO of Sony Computer Entertainment) and then-Sony CEO Mr. Idei. FeliCa Network exploited the jointly developed mobile e-commerce platform, and DoCoMo came to offer full-scale mobile e-commerce.

In the broadcasting domain, the joint venture formed exploratory networks with leading broadcasters, targeting a service that fused mobile phone connection and terrestrial digital broadcasting. In the content and the Internet domain, it established strategic cooperation and joint ventures, and dynamically formed exploratory networks with internal and external development partners (such as the US company Texas Instruments) aimed at joint development of core technology (hardware and software) for mobile phones.

Given the short life cycles of mobile phone development and the high functionality required of its handsets, the development of the central LSI system (Kodama and Ohira 2005) through links with semiconductor companies was urgent business,. DoCoMo shared its mobile phone development roadmap with handset development manufacturers and semiconductor manufacturers, and entered the market for mobile handsets loaded with timely and rhythmical new functions. With the i-mode mobile phone business creating successive waves of S-curves, creative dialogue was promoted to form a new environment (structure).

With these exploratory networks, various problems and challenges were dialectically synthesized, i-mode and roaming services were realized overseas (specifically through i-mode licensing contracts with communications carriers in Germany, the Netherlands, Taiwan, Belgium, France, Spain, Italy and the United States, and the launch of i-mode services abroad), and the "mobile wallet" was created as a key lifestyle tool service. "Mobile wallet" services enable the acquisition of shopping, all kinds of coupons, and members' cards with electronic money and mobile credit.

Recently, moreover, it has become possible to exploit the mobile phone to verify employee status as a step in corporate security management. Further expansion into the areas of global scale, different industries, and expanded technologies are major features in phase three exploratory networks as distinct from those in phase two. It has also become possible to develop and provide mobile handsets as "lifestyle tools" loaded with various new functions (ubiquitous communications functions beyond phone and email) (see Fig. 4.13). By May 2006, DoCoMo had 46.70 million i-mode subscribers and 25.36 million FOMA subscribers.

Through the practice of providing new services in various locations, DoCoMo employees at stage three have acquired and accumulated still higher quality and more diverse knowledge. In the third phase, under the restrictive conditions of domestic market saturation, DoCoMo planned to switch from a domestic to an international focus, migrate to a new generation system, and create a mobile phone culture with new lifestyle tools. In this phase DoCoMo has, through practice, created new overseas and "lifestyle tool" markets (structures) from dual networks—more exploratory networks and exploitative networks. Meanwhile the newly created structures are promoting further competition with rivals, and leading to far more new innovations and strategies in organizational networks centered on DoCoMo (see Mobile Phone Phase 3 in Fig. 4.3).

The kind of behavior described above is not exclusive to DoCoMo. The au (KDDI) and Vodafone mobile phone carriers are also simultaneously working to create new markets and establish and disseminate existing markets through the activity of dual networks (exploratory practice from exploratory networks and exploitative practice from exploitative networks).

From the viewpoint of the "corporate innovation streams" mentioned in Chap. 2, moreover, DoCoMo has created new markets and competitive environments by implementing an "environment creation strategy" cultivating new mobile phone markets as new lifestyle tools, and further implemented a new "environment adaptive strategy" adapted to the new competitive environment.

4.3 Forming Small-World Networks Within the Company

At the time of its split from major player NTT in 1992, NTT DoCoMo was a small organization of around 2,000 people. At first DoCoMo employees were transferred from NTT divisions and contracted from outside companies, and in the melee of strangers thrown together as work colleagues, a management style arose where leaders at management level from each functional organization (divisional and departmental heads) would all participate without regard for rigid, traditional hierarchical thinking. This gave birth to a consensual culture and climate built up from thorough debate among leaders from each functional organization division, with each arguing their corners in the same arena. The result was the formation of cross-functional project teams that were both flexible and adaptable.

DoCoMo's top and middle management layers shared a corporate vision of quickly establishing a Japanese mobile phone market and building a new mobile communications environment, distinct from Japan's public fixed phone services, where communication could be enjoyed by "anyone, anywhere, anytime." To rapidly expand the new mobile phone market on the basis of this vision, top and middle managers dynamically implemented swift decisions made through sharing information and knowledge at each management level and engaging in exhaustive debate around mobile communications infrastructure management and the expansion of new services (developing new mobile phone terminals and services). In this writing I will refer to the small-world networks (SWN) formed from leaders and managers at each management level and crossing organizational boundaries as "leadership teams" (LT).

In July 1988, leadership teams were split into top and middle management layers to absorb the expansion of organizational scale accompanying the growth (in terms of both earnings and scale) of the DoCoMo corporation and its mobile phone market. To strengthen links between top- and mid-level leadership teams, top management LTs with middle management representation exhaustively debated important topics proposed by middle management LTs before top management executed prompt and appropriate decisions.

At its launch as a new company, DoCoMo set out a corporate vision of its basic values, behavioural standards, and "corporate philosophy." At the core of this concept is the thinking that mobile communications liberate people from the confines of time and space, and act as a tool enabling access to any time and place. This mobile communication engenders new business styles and lifestyles, and begins to create a new communications culture unprecedented in terms of time, space, and form. Put another way, the mobile communications services trigger the development of a new type of knowledge society. DoCoMo had a major role to play in realizing this society, and energetically took on issues such as rich service functions, expansion of service areas, development of more detailed, higher-quality services, and setting of low fees aimed at enabling any user to access mobile communications anywhere. Many of the people who transferred from NTT at the time of the split maintained and transferred the positive elements of the traditional organizational capabilities (such as a high-quality, disciplined business system in a disciplined organizational culture) from their time at NTT, including research and development, network design and construction, high-speed network operations technology, and a nationwide sales force, while at the same time nurturing a creative, youthful corporate culture capable of maximizing the capabilities of all its employees. From the time of start-up, DoCoMo became a symbiotic organizational structure comprising functional organizations, which possessed assets that had been inevitably hidden during the NTT era, and project-based organizations, which possessed a new corporate culture (quite different from that of the NTT era) that later became the driving force behind new marketing innovations such as i-mode development. These leadership teams as SWNs became dual (exploratory and exploitative) network nodules. In other words, they were positioned as hubs or nodes in a network (see Fig. 4.6).

4.4 Network Integrative Competences Through Leadership Teams

As mentioned above, DoCoMo's corporate activities are supported by a clear twenty-first century vision of pursuing long-term positions and creating absolute value by realizing a new communications culture and knowledge society through mobile technology. Continuous knowledge-creating activities take place aimed at realizing the corporate mission and business domain based on this vision. Key DoCoMo characteristics include continuous innovation and the dialectical synthesis of project-based organizations possessing new, heterogeneous knowledge assets with functional organizations possessing expertise accumulated over many years as knowledge assets.

Project-based organizations are creating new business model concepts (including new products, services, and business frameworks) through trial and error, founded on imagination and creativity and aimed at innovation in an uncertain environment. They are forming multiple exploratory (project) networks with external strategic business partners and implementing emergent strategies (Mintzberg 1978; Mintzberg and Walters 1985). Individual projects within project-based organizations take independent distributive action (Nonaka and Toyama 2002) as network organizations. Business directions and objectives are regulated for the whole body of project-based organizations, however, and their business activities constantly monitored from the organization's top level.

These project-based organizations constantly lead to the creation of concepts, prototypes, and incubations for new products and services. The business processes of equipment assembly, sales, distribution, aftercare, and support are important to promptly and efficiently invest in, distribute and expand new products and services. The burden of these business processes is carried by the infrastructure of the functional organizations, which drive practice and processes as bureaucratic organizations by forming traditional and "exploitative networks" (functional networks) with group companies, and strategic outsourcing partners based on knowledge assets accumulated over many years. Functional organizations draw up disciplined, deliberate planning strategies (Mintzberg 1978; Mintzberg and Walters 1985), but routines aimed at incrementally improving business process efficiency are daily occurrences. So the fruits of innovative new product and service concepts from project-based organizations are promptly, surely, and efficiently invested in the market before being distributed and expanded.

These two types of organization (project-based versus functional) and network (exploratory versus exploitative) are general classifications. They possess paradoxical elements such as creativity and independence on the one hand, and efficiency and tradition on the other, and disagreement and conflict commonly break out between the organizations. These elements are a major hindrance to the integration of actors' knowledge in each organization. It is the job of the SWN leadership teams to promote this integration. Leadership teams comprise leaders (CEO, executives, division heads, departmental heads, and project leaders) and managers (project managers and assistant manager responsibilities) at each of DoCoMo's management levels, including top, middle, and mixed management teams at project-based and functional organizations, informal cross-functional teams, and task forces.

"Leadership teams" possess strong network connectivity to bind the dual networks. Specifically, leadership teams have the role of consolidating and synthesizing exploratory and exploitative networks—the two cohesive networks—through short cuts (not all networks are necessarily cohesive; networks of weak relationships also exist). "Network connectivity" is a connection capability of networks that realizes the gathering and synthesis of distinct creative and practical knowledge mentioned in Chap. 3. "Leadership teams" fuse and bind each piece of knowledge from the dual networks, and creatively and systematically establish the strategic methodology of setting up coexisting contradictions of opinion, giving rise to DoCoMo's general "network integrative competences" (see Fig. 4.6).

With leadership teams, each leader must display dialectical leadership (Kodama 2007a) focused on DoCoMo's vision and corporate mission. At the same time, leaders must thoroughly appreciate problems and issues through constructive debate, and expedite them through shared dialectical dialog. Each leader must also communicate and collaborate in order to recognize the role and value of each job. By so doing, leaders can engineer constructive outcomes from the various conflicts that arise among them. Meanwhile the CEO, as the final decision-maker in the leadership teams, must display top-down leadership when necessary, but must also strengthen communication links with individual leaders by actively creating time and space for dialog and debate within leadership teams, and maximize the coherence of each leader's dialectical leadership.

In this case study, strategy implementation is creating a spiral of unprecedented new markets by forming 'exploratory networks' (project networks) with content provider clients, strategic partners, and players in different industry fields, such as finance and distribution. Project-based organizations centered on the Products and Service Headquarter have launched successive new concepts aimed at realizing mobile phones as "lifestyle infrastructure," such as mobile e-commerce from handsets loaded with prepaid and credit cards and mobile phones loaded with digital terrestrial broadcasting functions.

The marketability of new services is first confirmed by moving through the processes of concept making, marketing, trial development, and incubation at project-based organizations. The distribution and embedding of these services in new markets that are swiftly commercialized by functional organizations (including equipment, maintenance, and sales divisions) is then promoted spirally. Functional organizations promote a series of efficient business process management cycles. These include efficient and accurate equipment investment planning policies that respond to predicted demand for new services, introduction of network operation systems to maintain high-quality services, nationwide sales and maintenance systems from exploitative networks (functional organization networks) resulting from strategic collaboration as well as outsourcing of group companies and sales branches, and the establishment of aftercare support systems.

DoCoMo's project-based and functional organizations discuss issues of emergent strategy and the planning that supports it, involving timing, type of strategy, tactics, mechanisms, and resources, and make decisions. This process takes place through leadership teams comprising teams of leaders from each project and functional organization (including marketing, research, service development sales, equipment, investment, or maintenance services) in each business. The leaders of these teams select strategies and tactics that could lead to the flowering of true marketing innovations after thorough dialog and discussion, and each leader moves to execute specific actions through dialectical leadership. The synergies of dialectical leadership through the collaboration of CEO and leaders from each management level, including executive staff, emphasize dialectical dialog, promote detailed, deliberate strategies for carefully selected emergent strategy issues, and realize the synthesis of knowledge and strategy from different networks. These "network integrative competences" launch spirals of new market (environment) creation and growth.

DoCoMo's network integrative competences, thus created by its leadership teams, resulted in the Tokyo launch, in May 2005, of the prototype FOMA as the world's first third-generation mobile phone service, and enabled the launch of full-scale services in October 2005. By the end of May 2006, DoCoMo had signed up 25.36 million subscribers.

From the perspective of the strategy practice process described above, DoCoMo's dynamic view of strategy around the nucleus of these leadership teams can also incorporate an integrative strategy arising from the coexistence of emergent and deliberate paradoxes (Kodama 2003).

4.5 The Dynamic Boundaries Congruence of Business Architecture

Finally I would like to consider the evolutionary process of DoCoMo's dynamic strategic management. In each phase of innovation DoCoMo managed to achieve congruence of the environment and corporate system, and of individual management elements within the corporate system (see Fig. 4.14). In the case studies mentioned above, DoCoMO implemented an "environment adaptive strategy" adapted to the constant environmental changes in the mobile phone markets in each of the three development phases. DoCoMo also set out an "environment creation strategy" that planned and implemented a business model to constantly create new markets at each development phase.

In the first phase, DoCoMo set out a new environment creation strategy with marketing and technology angles for the mobile phone market, where it had failed to make much headway, and built a pioneering, vertically integrated value chain linking the user, DoCoMo, communications equipment manufacturers (including mobile handset manufacturers), equipment assembly and maintenance companies, and sales companies. DoCoMo optimized vertical boundaries as a corporate system and dynamically achieved congruence with a changing environment. Later

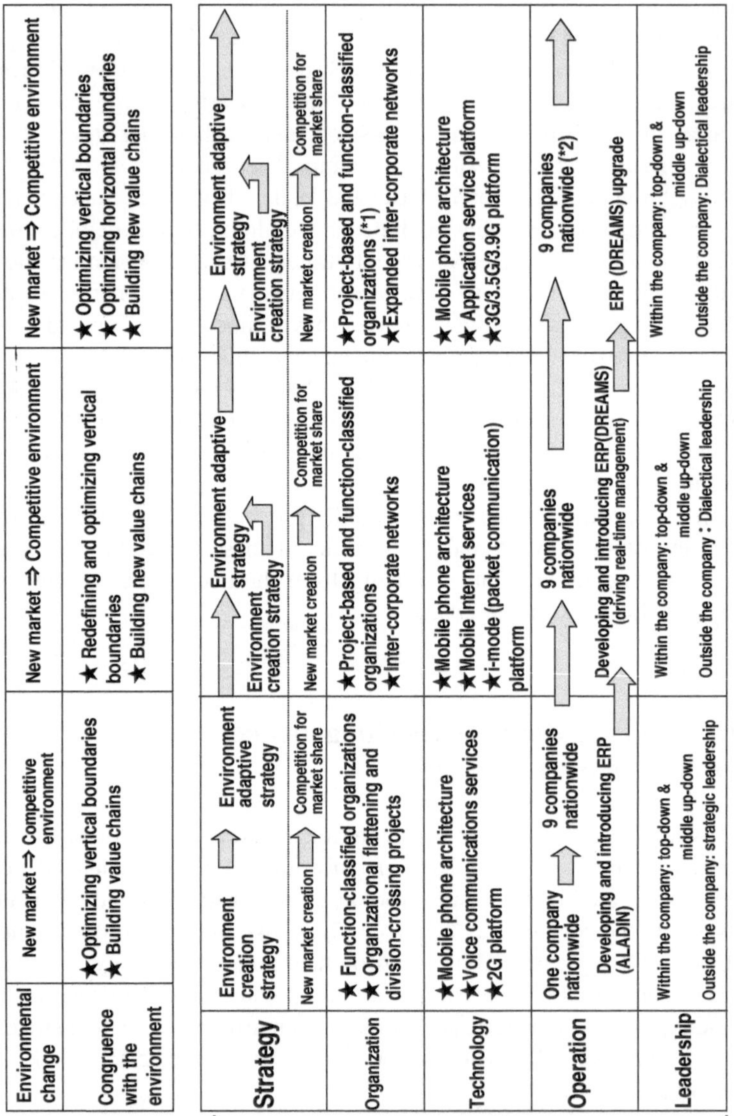

	Mobile phone phase 1	Mobile phone phase 2	Mobile phone phase 3
Environmental change	New market ⇒ Competitive environment	New market ⇒ Competitive environment	New market ⇒ Competitive environment
Congruence with the environment	★ Optimizing vertical boundaries ★ Building value chains	★ Redefining and optimizing vertical boundaries ★ Building new value chains	★ Optimizing vertical boundaries ★ Optimizing horizontal boundaries ★ Building new value chains
Strategy	Environment creation strategy → Environment adaptive strategy New market creation → Competition for market share	Environment adaptive strategy ← Environment creation strategy New market creation → Competition for market share	Environment adaptive strategy ← Environment creation strategy New market creation → Competition for market share
Organization	★ Function-classified organizations ★ Organizational flattening and division-crossing projects	★ Project-based and function-classified organizations ★ Inter-corporate networks	★ Project-based and function-classified organizations (*1) ★ Expanded inter-corporate networks
Technology	★ Mobile phone architecture ★ Voice communications services ★ 2G platform	★ Mobile phone architecture ★ Mobile Internet services ★ i-mode (packet communication) platform	★ Mobile phone architecture ★ Application service platform ★ 3G/3.5G/3.9G platform
Operation	One company nationwide → 9 companies nationwide Developing and introducing ERP (ALADIN)	9 companies nationwide Developing and introducing ERP(DREAMS) (driving real-time management)	9 companies nationwide (*2) ERP (DREAMS) upgrade
Leadership	Within the company: top-down & middle up-down Outside the company: strategic leadership	Within the company: top-down & middle up-down Outside the company : Dialectical leadership	Within the company: top-down & middle up-down Outside the company: Dialectical leadership

Congruence of the environment and the corporate system

Congruence among Elements within the corporate system

(*1) In July 2008 the HQ system was abolished and changed to a flatter organizational structure   (*2) In July 2008 a single company system replaced the 10 companies nationwide system

Fig. 4.14 Congruence with the environment and congruence among elements within the corporate system

DoCoMo set out an environment adaptive strategy to maintain market share in a changing competitive environment. To achieve congruence among management elements within the corporate system, DoCoMo implemented new strategies for business (including the measure of doing away with deposits), equipment (network equipment strengthening with risk), and technologies (mobile phone development from new technology architecture) supporting its environment creation strategy.

On the technology side, DoCoMo successively launched new mobile phone models to enrich customer services in a competitive environment and rapidly expanding mobile phone market. On the operations side, meanwhile, the organizational management of a single-company system for the whole of Japan instituted at DoCoMo's launch in 1992 was changed to a nine-company system (headquarters and regional subsidiaries) covering the whole country in 1993, preparing for a system capable of promptly providing detailed customer services in line with regional demands.

To cope with the increased workload accompanying the skyrocketing subscriber base and the launch of the sell-off service for phone handsets, DoCoMo launched the new ALADIN Enterprise Resource Planning (ERP) customer management system that integrated the databases distributed among individual business sections (subscriber reception, credit business, ROM writing, network construction, payment processing, and others), and rolled it out to the whole nine-company group in 1997. The installation of ALADIN, which possessed POS functions for the mobile phone business, brought about a dramatic improvement in business. To achieve optimization of these kinds of strategies, technologies, and operational interactions, DoCoMo needed to build a flattened, function-classified organizational system and division-crossing projects that would rapidly adapt to numerous challenges.

In phase one, both the strong leadership displayed by CEO Oboshi and middle up-down management centered on middle management (Nonaka 1988) were key elements of the DoCoMo leadership essential to implementing optimization among the individual strategy, organizational, technology, and operational elements. For external stakeholders, the strategic leadership was important for DoCoMo to display the vision and to guide stakeholders for creating new markets and building vertically integrated value chains.

Phase two created, disseminated, and established the new i-mode market out of the sense of crisis arising from the saturation of the voice communications market. Phase two differed from phase one in that DoCoMo built a pioneering vertically integrated value chain which redefined the corporate system's vertical boundaries and incorporated new stakeholders such as content providers. DoCoMo optimized vertical boundaries with the mobile Internet market, and then dynamically achieved congruence with the changing competitive market. Regarding congruence with the management elements within the corporate system, an aspect that differed greatly in the second phase was DoCoMo's emphasis on the philosophy of strategic versatility in order to implement an "environment creation strategy" realizing the various technology elements (new mobile phone architecture, mobile

Internet services, and an i-mode packet communications platform) required to create new markets and an "environment adaptive strategy" adapted to competitive environments.

Specifically, this involved establishing specialized organizations (such as i-mode business and mobile multimedia headquarters) implementing development from the planning and design of individual business domains and their constituent parts of individual business models, products, and services, and the building of numerous project-based organizations within the HQs. Then these project-based organizations synthesized the environment creation and adaptive strategies through close links with function-classified organizations dealing with areas such as sales, equipment, and maintenance. This phase placed greater emphasis on the formation of "exploratory networks" from inter-corporate networks with external partners.

On the operations side, DoCoMo drove business operations forward with a system of nine companies nationwide and the development and 2002 introduction of the new "DREAMS" intra-corporate information system (a radical update of phase 1's ALADIN system) targeting real-time management of business data, business innovation, enhanced managerial accounting, and a rapid response to systems accounting. As a result, it became possible for the whole DoCoMo group to acquire a real-time grasp of consolidated systems accounting data displaying the business condition of the whole group; business- and function-classified managerial accounting data displaying the performance of individual business and organizational roles; and operational data such as sales volume and traffic, thus speeding up business plan comparisons and subsequent actions.

In phase two, the leadership required to optimize the individual elements of these strategies, organizations, technologies, and operations came not only from DoCoMo's second president Keiji Tachikawa (Oboshi's successor), but also from middle up-down management centered on middle managers, which was an important aspect of DoCoMo's corporate culture inherited from phase one. Meanwhile, for external stakeholders, it became important for DoCoMo that not only the display of the vision to create the new i-mode market and build vertically integrated value chains, but also the dialectical leadership to form win-win relationship structures among the stakeholders.

The final phase, phase three, saw the creation and diffusion of new markets to establish a new post i-mode S-curve. In the new phase DoCoMo further optimized the vertical boundaries of the corporate system built in phase two while defining horizontal boundaries to advance into new business domains (a feature that distinguished it from phase 2). The mobile credit business has gained a sure foothold in the finance business, while businesses such as integrated broadcasting and communications, automobile-related telematics, and ubiquitous business will open up DoCoMo's horizontal boundaries and become triggers for setting out new business models from an "environment creation strategy." DoCoMo is now aiming to optimize vertical and horizontal boundaries under a changing competitive environment, and dynamically implementing congruence with the environment.

With regard to the congruence of management elements within the corporate system, DoCoMo is emphasizing still greater strategic diversity in the third phase. The aims are to implement an environment creation strategy that will realize the diverse technology elements (new mobile phone architecture, application service platforms, and 3G, 3.5G, 3.9G platforms) essential to creating new markets at vertical and horizontal boundaries, and an environment adaptive strategy adapted to competitive environments. The synthesis of environment creation and environment adaptive strategies through close linking of project-based organizations creating new business models and function-classified organizations including sales, equipment, and maintenance has been given greater emphasis from phase two on, and the formation of "exploratory networks" from inter-corporate networks among partners in different industries has received greater stress in phase three. In July 2008 DoCoMo reformed its organization to remove the bad practices that had arisen from organizational swelling of the HQ system in the second and third phases, ending the layered structure of the HQ system to create flatter divisional organizations retaining elements of the project-based organizations. By so doing, DoCoMo is planning to speed up in-house decision-making and strengthen links among divisions.

On the operations side, in July 2008 DoCoMo's business management system of nine companies nationwide reverted to the single-HQ system applied at its founding. DoCoMo's regional companies had migrated to the nine companies system in 1993, and had achieved consistent results from such aspects as development of sales policy models with close regional links and service area cover supporting local conditions. The group merged into one with the aim of enriching and strengthening customer services and achieving speedy, more efficient group management and so adapt to recent changes in the business environment surrounding the DoCoMo group. By implementing this new operational side, DoCoMo is realizing congruence with dynamically changing strategies, organizations, and technologies.

In phase three, the leadership required to optimize the individual elements of strategies, organizations, technologies, and operations came not only from DoCoMo's third president Masao Nakamura (Tachikawa's successor) but also from middle up-down management centered on middle managers, which was an important aspect of DoCoMo's corporate culture inherited from phase two. Meanwhile, for external stakeholders, it became important for DoCoMo that not only the optimization of i-mode's vertically integrated value chains, but also the dialectical leadership to display new business vision for horizontal boundaries and to form win-win relationship structures among the stakeholders.

At each stage of development, the key factors in achieving "boundaries congruence" lay in gathering and integrating practitioners' distinct creative and practical knowledge through the formation of SWS and networked SWS inside and outside the company (described in Chap. 3). These elements became the starting points for displaying dialectical leadership aimed at co-creating business models based on practitioners' collaborative leadership and new values, and building win-win relationships.

References

Ackoff, R. (1981). *Creating the Corporate Future*. New York: Wiley.

Banker, J. A. (1993). *Paradigms: The Business of Discovering the Future*. New York: Harper Business.

Barley, S. (1986). Technology as an occasion for structuring: evidence from observations of CT scanners and the social order of radiology departments. *Administrative Science Quarterly*, 31, 78–108.

Barley, S., Tolbert, S. (1997). Institutionalization and structuration: studying the links between action and institution. *Organization Studies*, 18(1), 93–117.

Christensen, C. M. (1997). *The Innovator's Dilemma: When New Technologies Cause Great Firms to Fail*. Boston, MA: Harvard Business School Press.

Dougherty, D. (1992). Interpretive barriers to successful product innovation in large firms. *Organization Science*, 3(2), 179–202.

Fleming, L., Sorenson, O. (2004). Science as map in technological search. *Strategic Management Journal*, 25(3), 909–928.

Giddens, A. (1984). *The Constitution of Society*. Berkeley, CA: University of California Press.

Hamel, G. (1996). Strategy as revolution. *Harvard Business Review*, 74(6), 69–82.

Hamel, G., Prahalad, C. K. (1989). Strategic intent. *Harvard Business Review*, 67(3), 63–76.

Hanna, T., Freeman, J. (1984). *Organizational Ecology*. Boston, MA: Harvard University Press.

Henderson, R., Clark, K. (1990). Architectural innovation: the reconfiguration of existing product technologies and the failure of established firms. *Administrative Science Quarterly*, 25(1), 9–30.

Johansson, F. (2004). *The Medici Effect*. Boston, MA: Harvard Business School Press.

Kim, W. C., Mauborgne, R. (2005). *Blue Ocean Strategy*. Boston, MA: Harvard Business School Publishing.

Kodama, M. (2003). Strategic innovation in traditional big business. *Organization Studies*, 24(2), 235–268.

Kodama, M. (2007a). *The Strategic Community-Based Firm*. Basingstoke: Palgrave Macmillan.

Kodama M., Ohira, H. (2005). Customer value creation through customer-as-innovator approach -case study of development of video processing LSI. *International Journal of Innovation and Learning*, 2(1), 175–185

Leonard-Barton, D. (1992). Core capabilities and core rigidities: a paradox in managing new product development. *Strategic Management Journal*, 13, 111–125.

Leonard-Barton, D. (1995). *Wellsprings of Knowledge: Building and Sustaining the Sources of Innovation*. Boston, MA: Harvard Business School Press.

Levitt, B., March, J. (1988). Organization learning. Scott, W., Blake, J. (eds.), *Annual Review of Sociology, Annual Reviews*. 14, 319–340.

March, J. (1991). Exploration and exploitation in organizational learning. *Organization Science*, 2(1), 71–78.

Markides, C. (1999). *All the Right Moves: A Guide to Crafting Breakthrough Strategy*. Boston, MA: Harvard Business School Publishing.

Martines, L., Kambil, A. (1999). Looking back and thinking ahead: effects of priorsuccess on managers' interpretations of new information technologies. *Academy of Management Journal*, 42(3), 652–661.

Mintzberg, H. (1978). Patterns in strategy formation. *Management Science*, 24(4), 934–948.

Mintzberg, H., Ahlstrand, B., Lampel, J. (1998). *Strategy Safari: A Guided Tour Through the Wilds of Strategic Management*. New York: The Tree Press.

Mintzberg, H., Walters, J. (1985). Of strategies deliberate and emergent. *Strategic Management Journal*, 6, 257–272.

Nelson, R., Winter, S. (1982). *An Evolutionary Theory of Economic Change*. Cambridge, MA: Belknap Press.

Nonaka, I. (1988) Toward middle-up-down management: accelerating information creation. *Sloan Management Review*, 29(3), 9–18.

Nonaka, I., Toyama, R. (2002). A firm as a dialectical being: towards a dynamic theory of a firm. *Industrial and Corporate Change*, 11(5), 995–1009.

O'Reilly, C., III, Tushman, M. (2004). The ambidextrous organization. *Harvard Business Review*, 82, 74–82, April.

Porter, M. (1985). *Competitive Advantage*. New York: Free Press.

Spender, C. (1990). *Industry Recipes: An Enquiry Into the Nature and Sources of Managerial Judgement*. Oxford: Basil Blackwell.

Tushman, M., Anderson, P. (1986). Technological discontinuities and organizational environments. *Administrative Science Quarterly*, 31, 439–465.

Weick, K. E. (1989). Theory construction as disciplined imagination. *Academy of Management Review*, 14(4), 516–531.

References

Chapter 5
New Knowledge Creation Through Leadership-Based Strategic Community

5.1 Introduction

The chapter provides new practical viewpoints in knowledge management and leadership theory of project management through an in-depth case study. It is argued that community leaders, particularly business community leaders, must recognize that a strategic community (SC) as "Small-World Structure (SWS)" comprises of diverse types of business and processes needed to achieve continuous business innovation. The community leaders serve an important function in creating a networked strategic communities (networked SWS).

The innovation of a telemedicine system in the field of veterinary medicine in Japan is taken as a case study. Here it is shown how a networked strategic communities (networked SWS) of business and customers has been used to develop a new Integrated Video Transmission System using ICT. In particular, it shows how community leaders have created networked SWS in which the university, hospitals, private businesses and non-profit organisations have worked together to advance virtual networking in the field of veterinary medicine.

5.2 Networks of Strategic Communities

Rapid progress in ICT is leading the way for gradual innovation in diverse areas including society, economy and industry. Ever increasing acceptance and use of Internet, Intranet is generating flatter corporations with novel and improved communications platforms as well as creating new-business model, for inter-corporation transactions as championed by e-commerce. The Internet platform is poised to significantly change work practices and process in corporate settings in supporting the lifestyle of individuals in their day-to-day living. Furthermore, it is stimulating the proliferation of small offices and home offices(SOHOs). New business styles based on such concepts as virtual teams and virtual community are representative of such a trend. (e.g., Bechard et al. 1996; Lipnack and Stamps 1997)

Amid such a change of the times, the advent of the new twenty-first century networking generations will usher in major changes in individual's value systems,

M. Kodama, *Boundary Management*, DOI 10.1007/978-3-642-03789-4_5,
© Springer-Verlag Berlin Heidelberg 2010

especially as they relate to living and working. At the same time, it is anticipated that increasing importance will accrue in the years to come to the manner of existence of, and new ways of thinking about, the "communities" represented by corporate entities and non-profit organizations, which constitute massive aggregates of individuals. In corporate settings, in particular, the knowledge management method, refined through rapid ICT sophistication, is being adopted to address internal particulars, bringing on structural renewals in affected areas (Hesselbein et al. 1998). But the important point here is that, no matter how ICT is taken in for business-handling innovation, a corporation's strategic behavior most importantly depends upon innovation of the value systems of the individuals concerned and of the knowledge accumulated by them. Leadership generative of business innovation on a continuous basis that strategically taps the knowledge of extra-corporate human resources including the customers will become important.

Toward such an end, it becomes most important for leaders of corporations to aggressively create strategic communities (SCs) as SWS tapping on their own organisations' as well as outside contacts, including customers, in leading positions for use in innovating their own in-house core knowledge while at the same time creating new values and offering them to their customers. (Kodama 2001, 2003)

SC is based on the concept of "ba" as a shared space for emerging relationships that serves as a foundation for knowledge creation (Nonaka and Konno 1998). "Ba" is an interaction space involving language and communication. Knowledge is created through the interactions among individuals or between individuals and their environment. Participating in a "ba" means transcending one's own limited perspective or boundary and contributing to a dynamic process of knowledge creation. In a SC, members including customers who possess different values and knowledge consciously and strategically create a "ba" in a shared context that is always changing. They continually create new knowledge and competencies as a new "ba" by merging and integrating a single "ba" or multiple numbers of "ba" both organically and from multiple points of view.

In this chapter, I define SC as both emergent and strategic, a collaborative, inter-organisational relationship that is negotiated and associated with creative yet strategic thinking and action in an ongoing communicative and collaborative process or involves several arrangements (e.g. strategic alliances, joint ventures, consortia, associations, and roundtables), which neither depends on market nor hierarchical mechanisms of control (Heide 1994; Lawrence et al. 1999). From the practical aspect, I see the SC as an informal strategic organisation possessing qualities with both a resource-based (or knowledge-based) view (Mintzberg et al. 1998) and a strategic view (Porter 1980). The resource-based view is an emergent, learning view of the community in a shared context, while the strategic view is a planning view that aims to establish a desired position in the target market.

SC is applied, for instance, in cases where enterprises are in a management environment beset by numerous uncertainties, where predictions are difficult, and management is searching for valid strategies. The task of SC is to emergently and strategically form and implement concrete business concepts and ideas. Trial and error such as incubation is necessary, however, and SC takes the stance that a strategy will emerge from among the collaborative actions. For the most part, middle management is at the center of the SC. They form informal and virtual teams both inside and outside the company, including customers, and actively generate emergent, entrepreneurial strategies and create new knowledge. They then produce new demand that did not exist before which in turn results in the emergence of a new technology and market (kodama 2003).

On the other hand, the acquisition of resources and transfer of knowledge between strategic partners is different from the creation of new knowledge. Knowledge creation occurs in the context of a community that is fluid and evolving rather than tightly bound or static. The canonical formal organization, with its bureaucratic rigidities, is a poor vehicle for learning. Sources of innovation do not reside exclusively inside firms but between them.

Accordingly, knowledge creation is an extremely important issue that sees knowledge as a property of communities of practice(Brown and Duguid 1991), "ba" (Nonaka and Konno 1998), communities of creation(Sawhney and Prandelli 2000), SC(Kodama 2001 2003; Storck and Patricia 2000), and networks of collaborating organizations (Powell and Brantley 1992), rather than as a resource that can be generated and possessed by individuals. When the knowledge base of an industry is both complex and expending, and the sources of expertise are broadly dispersed, the locus of innovation will be found in networks of inter-organizational learning rather than in individual organizations (Powell et al. 1996). Connection through the networks of SC based on the inter-organizational collaborative relationships is thus an important origin of knowledge creation, and new knowledge grows out of the sort of ongoing social interaction that occurs in ongoing collaboration between SCs.

In this chapter, I describe the development process in a big project in Japan that occurred over the past 6 years. This paper focuses on new knowledge creation process by synthesizing capability (Nonaka and Toyama 2002) through dialectical leadership in the networked SCs. An example of the establishment of networked SCs is the business case of innovation of a telemedicine system in the field of veterinary medicine in Japan that makes use of information and multimedia technologies.

The innovation of a telemedicine system in the field of veterinary medicine in Japan is taken as a case study. Here a networked strategic communities of business and customers developed new systems using ICT. It shows how community leaders have created networked strategic communities in which the university, hospitals, private businesses and non-profit organisations take part in the advancement of virtual networking in the field of veterinary medicine.

The case examines how the network of SCs involving the university, hospitals, the private sector and non-profit organisation implemented the new product development to create the world's first ever multimedia veterinary telemedicine system and the impact that the created networked has had on medical field from the standpoint of new knowledge creation process through the synthesis of various knowledge within the networked SCs.

In this in-depth case study, our analysis of knowledge creation focuses on the degree and process to which the networks of the SCs created new knowledge based on new technologies, social needs, customer needs and the various contents that were diffused beyond the boundaries of the SCs. The case is further analyzed from two standpoints described below.

The first angle analyzes the characteristics of SC networks that became the trigger causing the synthesis of knowledge diffused from the boundaries of individual SCs that were distributed from the three elements of their involvement in collaboration (DiMaggio and Powell 1983), embeddedness in collaboration (Granovetter 1985; Dacin et al. 1999)and resonance of values (kodama 2001) at which the SCs were formed.

The second angle discusses the synthesizing capability through dialectical leadership that the leadership-based strategic community comprising community leaders within networked SCs uses to dialectically synthesize the different knowledge of SCs that is distributed in the new knowledge creation process in a big project. Finally, this chapter discusses on the managerial implications as organizations make use of their dialectical leadership and innovations in their efforts to achieve new knowledge creation and innovation.

5.3 Summary of an In-depth Case

5.3.1 Current Status and Issues in the Field of Veterinary Medicine in Japan

In Japan at present, there are 27,000 veterinarians, about 30 percent of whom are working in practice and in researches related to public health inspection such as the inspection of food items. Recently new fatal infectious diseases such as the Ebola hemorrhagic fever are appearing on a global scale, and zoonotic infections (communicable to human and animals) contracted from animals and pathogenic microbes such as Escherichia coli O157 are crossing national boundaries so that the role of the veterinarian is expanding[1].

Several years later, health veterinary medicine departments in Japan were reorganized in several research institutes, and it has been pointed out that teaching staff numbers (totally about 320 professors, lecturers and teaching assistants etc.) at universities were about half those in Europe and the United States so that research and

---

[1]In recent years, veterinary medicine is in the process of advancing towards the prevention of various diseases with organ transplants and at gene level and explaining the phenomenon of life. These days, when veterinary medical treatment is being questioned on the basis of life ethics, from a wide-ranging viewpoint that integrates medicine, engineering and science, "the education of talented personnel and the establishment of research and development having the characteristics of veterinary medicine in the chemical view of data related to animal medicine" is essential for the mastering of veterinary medicine. The place where veterinary medicine differs from other natural sciences is that having discovered the meaning and value of the existence of all animals, it helps them to survive. With respect to the existence of industrial animals and companion animals, because of the existence of social meanings and values in each case, clinical veterinarians carry the burden of fulfilling that duty every day with a great sense of mission. (Dialogue with Dr. Hirose)

education levels were falling behind. At present, the research papers in veterinary medicine field published by Japanese researchers are about one third those in Europe and in the United States.

The European Union completed the standardization of common teaching criteria and qualifications for veterinarians in 1999, and the United States and Canada are also closer to internationalization by standardizing their national veterinary examinations and qualifications. These standards and qualifications are gradually being adopted in Japan because of its belief that "bringing veterinary medicine teaching levels close to international levels is a matter of urgency".

Unless the levels of education and research in veterinary medicine are brought in line with the European and the American standards, there is a serious fear that only Japan's veterinary medicine will be left out from international veterinary medicine arena. Under these conditions, international cooperation and information sharing between veterinarians in the future is absolutely essential. This important issue is an issue common to industry, academia and government, an issue that must be solved jointly.

In January 1996, Dr. Hirose[2], Professor Emeritus of Obihiro University of Agriculture and Veterinary Medicine, built the "Animal Medical Information and Science Development Research Institute" (situated in Obihiro, Hokkaido, hereafter called the institute) as a joint venture of industry and academia. The research staff who constitute the Institute consist of people from different working backgrounds such as university instructors, practising veterinarians, group veterinarians, businessmen, dairy proprietors, and general citizens. In his support of the joint venture thas is what Dr Hirose had to say during my dialogue with him:

> Giving greater significance to veterinary medicine in the community not only requires the gathering of the opinions of us veterinarians who are specializing in veterinary medicine but we must also prevent the narrow fixing of ideas by creating the opportunity to open up experience and ideas by having the participation of many people in other fields (different industries). In particular, when it comes to the prevention of disease to minimize economic losses with respect to production livestock, we can never expect true results without a practical plan that involves the participation of the producers (agriculturalists). In other words, it is not enough to have a veterinarian with abundant knowledge. Several veterinarians who can pool their knowledge will be required in order to dig up the true values. Now is the time when our eyes must be turned towards what the producers are earnestly hoping for, the activities of veterinarians with such flexible ideas as this. It is very important to exercise

---

[2] Dr. Hirose is the Director of the Animal Medical Information and Science Development Research Institute and Professor Emeritus of Obihiro University of Agriculture and Veterinary Medicine. He specializes in clinical veterinary radiology. Using Japan's first imaging diagnostic terminology, he is a world pioneer in contributing to the development of imaging diagnosis by developing and introducing large animal X-ray diagnostic vehicles and industrial animal general imaging diagnostic vehicles, which even now he is active in bringing into the farmyard. He was awarded the Japan Veterinary Association President's prize in 1972. He was a director of the International Veterinary Radiology Association and sits on the review board of the Veterinary Radiology and Ultrasound Journal. He is the chair of the organizing committee for the 12th International Veterinary Radiology Association (IVRA) to be held in Obihiro, Hokkaido, in August 2000. See http://www.obihiro.ac.jp/~ivra/

free thinking in order to break down inflexible veterinarians. If the conventional framework is not demolished when it needs to be, there will be no new ideas or new personnel.

For interdisciplinary research, it is certainly necessary to step out a framework in the area of modern clinical veterinary medicine and to create an interdisciplinary organization from previously non-existent collective bodies (collective bodies of different industries). There should be collaboration in creating together within the collective body, with each veterinarian developing their own original ideas. (Dialogue with Dr. Hirose)

The aim of the Institute was to construct a virtual organization with 54 research members scattered throughout Japan, "a concept that gives full rein to independence in towns and villages, that promotes local expectations and characteristics and is not centralized in Tokyo, the capital of Japan." An additional objective, from the international viewpoint, was that it would serve as a base for the transmission of information. Meanwhile, the preparation of a new media environment in the area of clinical veterinary medicine can never be said to have progressed. Utilizing new media for education and research in veterinary medicine and the activation of local communities were some of the key points. Specifically, interactive communications and information sharing using a digital-based videophone greatly broadens communication among veterinarians and facilitates the expansion of the interactive telemedicine network.

5.3.2 Formation of Strategic Community and Networked Strategic Communities

For some time past, X-rays, Computed Tomography(CT), Magnetic Resonance Imaging (MRI), and so on have been used in veterinary medicine on both large animals (such as cattle and horses) and on small animals (such as dogs and cats) and are still being used for imaging diagnosis of various types of disease with great success in early treatment and prevention. In particular, the latest trends, that is to say "the enhancement of the status of companion animals in a household as people age and have fewer children," are creating a major flow, in veterinary medicine as in the human world, "from treatment to prevention" and the weight of imaging diagnosis is growing.

Amongst these moves, research workers such as Dr. Hirose, a central figure as an innovative customer, have introduced the first X-ray diagnostic vehicle in Japan dedicated to large animals to all universities in Japan and have been conducting a great number of pioneering trials (Hirose 1985). Nevertheless, the current picture is that the history of imaging diagnosis in veterinary medicine is still shallow and there are few specialists. For that reason, it has become necessary for distant veterinarians to collaborate in education, diagnosis and instructions for treatment.

On the other hand, lately the fence between the diseases of animals such as pets and humans is low and, as a result, for the veterinarian, the time has arrived when quick action by imaging diagnosis is demanded. On that point, the demand for bi-directional remote treatment by imagery is growing stronger, not only simply between veterinarians but also in areas of clinical veterinary medicine such as "between veterinarian and large animal producer" and "between veterinarian and small animal breeder." On this basis, at the start of 1996, the research group centered on Dr. Hirose and the Nippon Telegraph and Telephone Corporation (NTT) project group began research and development of remote diagnosis in veterinary medicine.

The NTT project group was faced with the immediate problem of how to popularize a desktop video conferencing system that they had started to market in March 1996. One of the project targets was to develop the use of an application using the video conferencing system in the field of medical treatment. The NTT project leader's problem was how to approach customers with the new system that had been developed and have them appreciate its commercial value. However, in 1996 there were few customers in Japan who needed video conferencing with personal computers and the necessity to use a video conferencing system was not deeply rooted in the culture of the entire community. However, the vision and idea of the NTT project leader was that the concept of "creating a new interactive video communication culture in Japan" should definitely be achieved. Expediting its use in the fields of welfare of the elderly and education, and especially in all medical care including the field of veterinary medicine, was an important issue. In a dialogue with NTT project leader, this is what he said:

> I thought that the key to popularizing the video conferencing system, which is a tool for interactive video communication, was a hint to research on the form of usage by customers. Video conferencing systems are generally used for virtual conferencing in business, but in analyzing the future market, he inferred that the demand in telemedicine and distance learning would grow. However, we are a telecom carrier, and when it comes to medicine and education, we had not built up any knowledge, we had absolutely no know-how, and we had much to learn from well-versed doctors and teachers in these areas. We believed it was quite possible that the field of medical treatment, from the medical care of humans to that of animals, would be a potential area for the application of video conferencing systems and that a partnership with Dr. Hirose, who is an authority in the area of clinical veterinary medicine using radiology, would be an important business opportunity. At the same time, from the viewpoint of animal protection, we thought that the fact that the Phoenix which we had developed could contribute to the development of clinical veterinary medicine was a matter of very deep significance (Dialogue with NTT project leader).

Dr. Hirose's research group planned to construct their own information exchange system using digital access networks in the area of veterinary clinical medicine. This system was constructed on the basis of summarizing the opinions of the Institute's 54 participating research members. The main objective was for all of them to be able to exchange clinical image information easily on the same level. Specifically, it was a system structured to suit the mutual exchange of information with a uniform method, in which all research members would have the same equipment. Commenting on the above, Dr. Hirose said:

> For us veterinarians to develop our own telemedicine system in the area of clinical veterinary medicine would not be possible with only our own knowledge and know-how. A tie-up with NTT, with their many years of knowledge and know-how in communication network technology and image communication technology, was a splendid opportunity to

produce new innovation through collaboration that concentrated the experts of industry and academia in partnership (Dialogue with Dr. Hirose).

In this way, the visions and values of both the NTT project leader and Dr. Hirose, who was a customer, coincided, and a partnership for reaching the one single objective was established. The enterprise and customers formed a SC (SC-a in Figs. 5.1 and 5.2) consisting of a virtual team with about seven project members from NTT, and the customers were the academic members in Obihiro University and the veterinarians group of 54 (scattered geographically throughout Japan) with Dr. Hirose at the Institute(See Phase 1 in Fig. 5.2). Then all members who formed the SC-a were brought into resonance with values based on the same vision and concept, becoming as one body fired with enthusiasm for creating fresh innovation in the area of clinical veterinarian medicine. Furthermore, in order that the SC-a that they had formed would be maintained continuously, it was important for Dr. Hirose and the NTT project leader to arrange for the continuous application of empowerment and motivation with respect to the community members.

NTT proceeded to develop a new method of use for the video conferencing system by incorporating into their own company product the knowledge possessed by Dr. Hirose and his group, which was rooted in their many years of experience. Then, thanks to innovation from the knowledge within the SC-a, over a period of about one and a half years they developed a new telemedicine system and its popularity proceeded to spread throughout Japan. (In this chapter, the leader of the business that

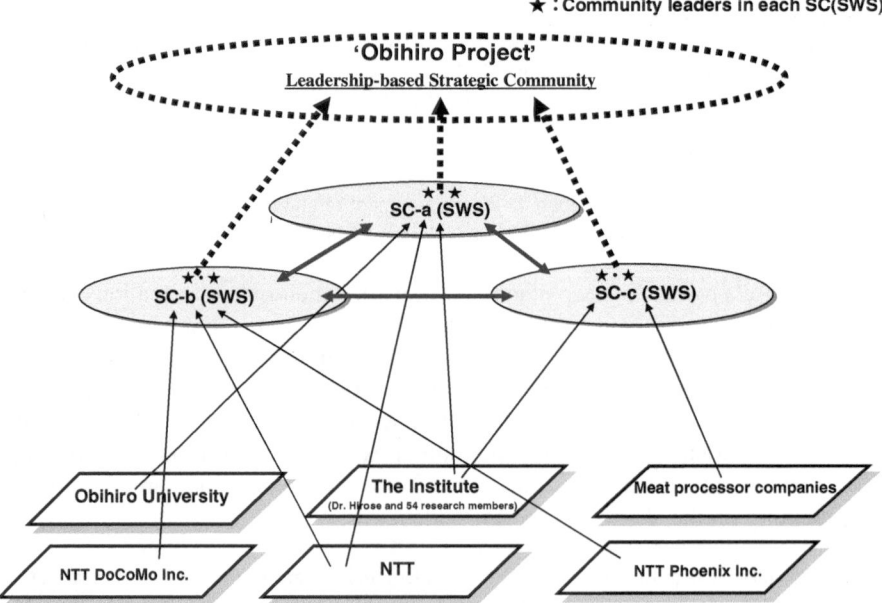

Fig. 5.1 Formation of strategic community and networked SCs

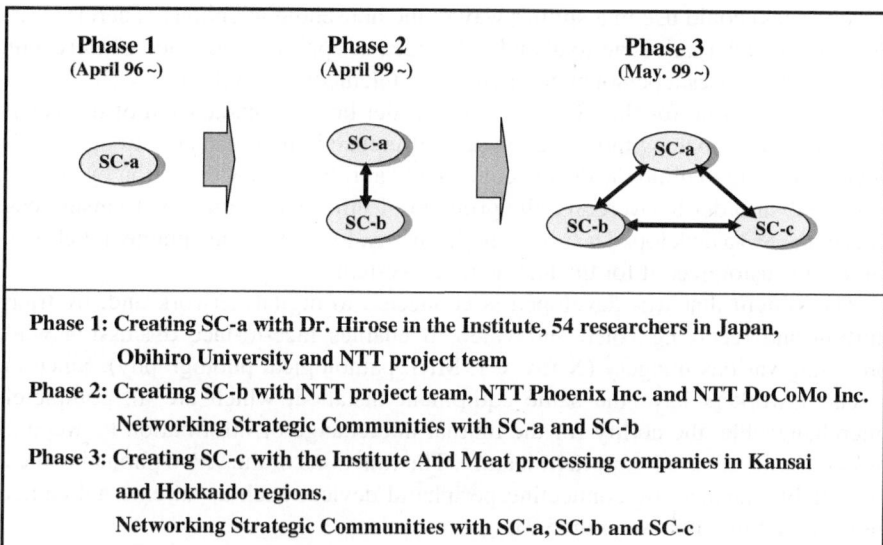

Fig. 5.2 Networking strategic communities at the big project

becomes the core, playing the central role within the SC of business and customers is called the community leader).

The biggest issue of the SC-a was that of deciding what would be the ideal telemedicine system for veterinarians. The assessment of the system would not only be based on performance and cost, but also easily use, it was necessary to create a system that many veterinarians would find easy to use. The issue for the SC-a at the time of starting up was the conception and design of the telemedicine system. The information and knowledge possessed by community members from different industries was shared and the dialectical dialogue between the many community members towards the design of the ideal telemedicine system triggered mutual knowledge within the SC-a. The factor that bonded each member to the SC-a was the common value with respect to their vision and concept of "innovation in the field of veterinary medicine." In order to create and provide new values for many veterinarians, while bearing in mind what is meant by "the application known as interactive video communication," new knowledge was triggered and created by both the physical and intangible aspects of the telemedicine system, including its conception, design, construction and method of utilization.

However, a major problem occurred at the initial startup stage. The problem arouse from the use of personal computers by veterinarians. Although a personal computer proved to be an extremely convenient device, it could not serve as an excellent data system. Instead of matching the person to the personal computer, since the informationally-handicapped would not come forward, it was necessary to construct a system equipped with the ideas and characteristics of the SC-a and having functions for the exchange of information that anyone (including veterinarians of various ages) could use in a similar way to the household telephone. Therefore, the SC-a had no alternative but to abandon the idea of applying the video conferencing system, which uses a personal computer, as a telemedicine system.

The next hurdle for the SC-a was to consider how to enable each of the veterinarians to participate without feeling any difference in their capacity to operate the device. Bearing in mind the design concept of "ability for easy operation by any veterinarian" in order to overcome the problem of information literacy, to insure ease of use, the SC-a developed a simple, high-quality, low-cost video phone in February 1998 and customized it for the telemedicine system.

The system that was developed is connected to digital network and, by transmitting and receiving voice and video, it enables face-to-face discussion while providing various imagery (X-ray, CT, MRI, pathological photography). Since the research workers have the same equipment installed, which are all completely interchangeable, the ability for the mutual interchange of information between all research workers is an important feature. The functions of the video phone itself can be greatly enhanced by connecting peripheral devices such as an external camera and external monitor (see Fig. 5.3).

The system is cost-effective in their one complete set cost around 400,000 yen. Moreover, a system that can be accessed by all veterinarians throughout the country has been developed for successively storing the various types of image data, such as cases and instances that are transmitted in real-time, in video-on-demand (VOD) (kodama 1999). In this way, comparatively low-cost technology has

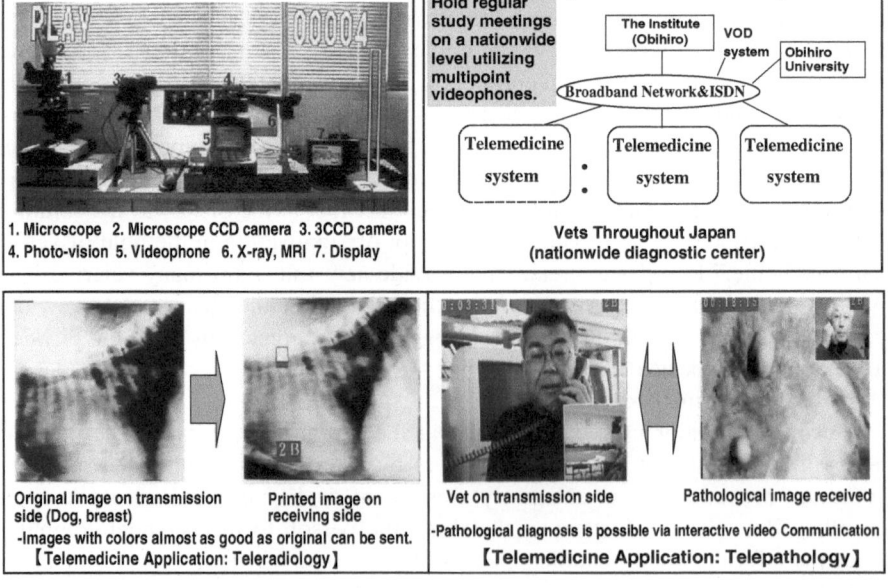

Fig. 5.3 Telemedicine system that was developed and its manner of use

been established that can transmit high-definition images of various case types to facilitate remote diagnosis and can also be used in the field of veterinary medicine.

Meanwhile, NTT was geaning up toward the further advanced new product development. The NTT project team was aiming at a practical integrated video transmission multipoint service that would connect seamlessly between Integrated Service Digital Network (ISDN) circuits, the Broadband Internet and the mobile network (the network that is the base for mobile phones and personal handy phones). Below the comment made by NTT project leader during a dialogue with him:

> Our target was to make the new telemedicine system practical. What we can expect is that by interconnecting with the 3G mobile network (IMT-2000)[3], which is scheduled to start services from March 2001, and digital ISDN networks and the Broadband Internet, customers anywhere, can have seamless interactive video communication at any time with anyone without being aware of the type of communications network. Then I believe that video phones will be used not only for business but their use will be extended to various fields such as education, medical treatment and welfare as well (Dialogue with NTT Project leader).

Therefore NTT project team was able to form the SC in April 1999 (SC-b in Fig. 5.1) with NTT Phoenix Inc.(the largest multipoint videoconferencing connection service company) and NTT DoCoMo Inc.(the largest mobile telephone company in Japan), to develop and realize the Integrated Video Transmission System (See Fig. 5.4). The system was set up to interconnected between the Institute the 54 veterinarians groups, university and hospitals through the video transmission by video phones, PHS (Personal Handy Phone) and 3G mobile video phones (See Fig. 5.4). A variety of technical issues related to video communication and transmission of educational and medical contents between SC-a and SC-b were thoroughly discussed. (See Phase 2 in Fig. 5.2)

To progress the formation of networked SC-a and SC-b, Dr. Hirose and NTT project leader made a presentation of the vision and concept for the project to the executive director and senior officials of the Ministry of Posts and Telecommunications(MPT). The aim was to gain the financial support for this project. As of March 1999, "Image transfer technology by Integrated Video Transmission Systems" proposed by Dr. Hirose and NTT project leader has been adopted by the "Fiscal 1998 new big development project" of the Telecommunications Advancement Organization of Japan (hereafter TAO),[4] which

---

[3]Current mobile phones and portable terminals are capable of high-speed transmission, but these are next-generation mobile, portable video terminals that can transmit video, and not just voice and text, interactively. W-CDMA and cdma2000 are next-generation mobile telephone systems proposed by the Swedish company Ericsson and the U.S. company Qualcomm, respectively. In Japan, FOMA (Freedom Of Mobile Multimedia Access) is the service brand name for DoCoMo's third-generation mobile communications system (IMT-2000). The FOMA market began to experience rapid growth from December 2002, reaching 1,500,000 subscribers as of October 2003.

[4]The Telecommunications Advancement Organization of Japan is a corporation approved by the Ministry of Posts and Telecommunications that provides various types of assistance in respect of research and development and telecommunications advancement projects in the field of data communication. See http://www.shiba.tao.go.jp/

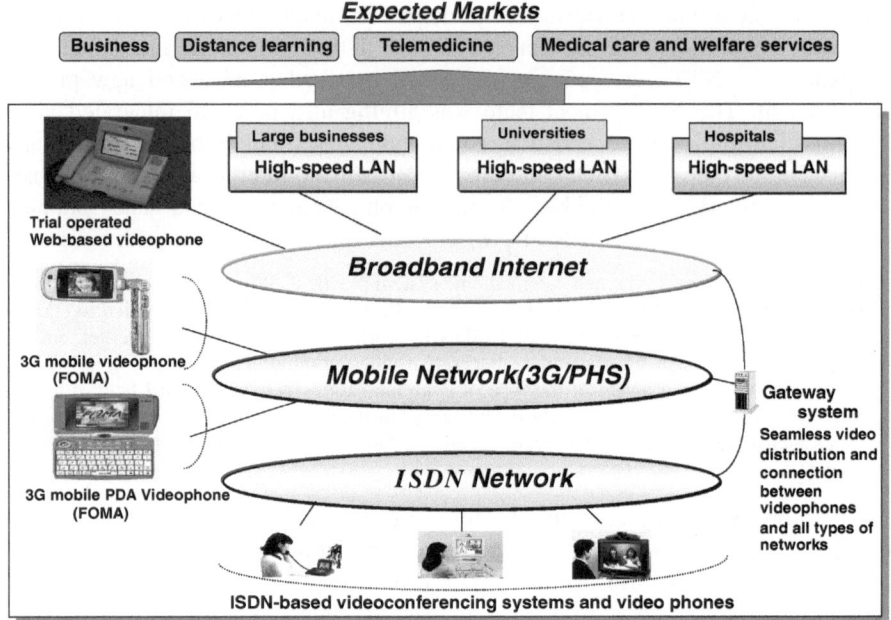

Fig. 5.4 Integrated video transmission systems

is a corporation approved by the Ministry of Posts and Telecommunications (MPT), and eligible for the investment of research and development funds by the Japanese government (Tokachi Mainichi News 1999a, 1999c). The Institute, NTT, NTT Phoenix Inc. and NTT DoCoMo Inc. could obtain research funds from the Japanese government and began to develop the new telemedicine system on the Integrated Video Transmission Systems.

New knowledge concerning the development of the new telemedicine system using new technology, and the development of methods for using the system, was shared among community members. The community members in SC-b were planning to acquire new image transmission technology in order to achieve high quality interactive video communication on the Integrated Video Transmission Systems. At the same time, Dr. Hirose and other community members were aiming at research on the form of usage. As a specific field trial, they thought of connecting the Institute in Hokkaido and a meat processor in the Kansai region and other local area in Hokkaido with the Integrated Video Transmission Systems and conducting a remote pathological examination of a large meat animal (beef cattle, for instance) (they formed SC-c in Fig. 5.1). Accordingly, an umbrella networked project centered on the Institute in SC-a, which was linked SC-a through SC-c in, was formed. This would coordinate the interworkings of different entities toward the realization of the high quality interactive video communication on the Integrated Video Transmission Systems, and a decision was taken to work out the details (See Phase 3 in Fig. 5.2). A proposal based on a vision and concept developed by Dr. Hirose and an NTT project leader triggered the realization of a community as an executive-created project, which in turn provided a vehicle for the mutual sharing of values held by the core leaders.

In April, 1999, an organism named Obihiro Veterinary Medicine Promotion Project ("Obihiro Project" hereafter) was formed that worked across various organizations, such as University, hospitals, Research Lab. as an NPO, NTT, NTT Phoenix Inc. and NTT DoCoMo Inc. (See Fig. 5.1). The purpose was to take further steps toward the accomplishment of the project. The "Obihiro Project", which was institutionalized among the core leaders in each of the various organizations (In this chapter, the "Obihiro Project" will be referred to as "Leadership-based Strategic Community; LSC"), was developed to aggressively promote new product development regarding the Integrated Video Transmission Systems. The core leaders of the various organizations (enterprise, universities, hospitals, medical associations, NPOs, etc.) will be referred to as "community leaders." If this research and development implemented by LSC is successful, it will make it possible to conduct a reliable virtual form of prior examination by a specialist veterinarian with respect to various communicable diseases (such as mad cow disease and O-157) before processing a large animal for meat.

5.3.3 Innovation in New Knowledge Creation

In the space of about a year, the LSC succeeded in the trial manufacture of the latest type of web-based video phone,[5] mobile streaming system and mobile gateway system compatible with the Transmission Control Protocol/Internet Protocol (TCP/IP) and Third Generation (3G) video transmission protocol and began to experiment from August 2001 (Ohira et al. 2003). The LSC used a web-based video phone and 3G mobile video phone and conducted various image transmission experiments by connecting with the mobile gateway system to three types of networks: the Broadband Internet, an ISDN circuit, the 3G mobile network including Personal Handyphone Subscriber (PHS) network, and proved the potential practicality of the new telemedicine system in the area of veterinary medicine. Then, with the aim of making the "telemedicine system," (which was potentially usable on all of these networks) fit for practical use and of establishing the use of this new system in fields that should be called "remote radiology" and "remote pathology," Dr. Hirose's team promoted research aimed at pioneering the form of the next generation of medical treatment. Meanwhile, as a telecom carrier, the NTT project team was aiming at a practical integrated video transmission service that would connect seamlessly between various telecommunications networks. Then, with this experiment, they decided on a strategy of virtual business using video phones including mobile videophones in the fields of "distance learning," "telemedicine," "welfare" and "remote relay" (Nikkei Communication 1999; Nippon Kogyo Shimbun 2000).

---

[5] A movable video phone that uses the Internet communication protocol TCP/IP and is capable of accessing the Web.

5.4 Results and Discussion

In this section, I will use six dimensions derived from the research method to consider what sort of impact the SC networks and the synthesizing capability through dialectical leadership of the community leaders had on new knowledge creation and the output of "Obihiro Project." As a first angle, I analyze the characteristics of SC networks that became the trigger causing the synthesis of knowledge diffused from the boundaries of individual SCs that were distributed from the three dimensions of their involvement in collaboration, embeddedness in collaboration and resonance of values at which the SCs were formed. As a second angle, I discuss the synthesizing capability through dialectical leadership that the leadership-based strategic community comprising community leaders within networked SCs uses to dialectically synthesize the different knowledge of SCs that is distributed in the new knowledge creation process in this big project.

5.4.1 Characteristics of Networked SCs

Information and knowledge within communities is both sticky and leaky. It is important, however, that the networked SCs makes the leaky aspect of knowledge in communities work in a positive manner and that community leaders of all SCs promote the sharing and inspiration of knowledge beyond the SCs' boundaries. The act of transcending boundaries stimulates deep, meaningful learning, which in turn opens possibilities for the generation of new knowledge and creativity. Radically new insights and developments, such as the big project in this case, often arise at the boundaries between SCs (Grant and Baden-Fuller 1995; Grant 1996). In particular the dilemma faced by organizations is the need to reconcile rapid access and synthesis of relevant new knowledge, with the long time frames needed for knowledge creation and synthesis. Networked SCs based on deep inter-organization collaboration can offer a possible solution. The need for businesses and society to transcend the boundaries of SCs has been increasing in recent years as markets in their industry become more uncertain and technology as well as customers' needs change more rapidly. One practical experience and the knowledge of a single organization need not to be embraced in order to solve the complex sort of problem of synthesizing technology as well as social needs that is shown in this case. The community leaders of organizations need to actively build SCs inside and outside the organization, including with customers, and then to transcend the boundaries of the SCs and network them in a speedy manner.

One important element in the formation of networked SCs is collaboration. The presence of deep interaction among community members, the presence of a strategic partnership among organizations that form networked SCs, and interactive information sharing within networked SCs are not the only conditions required for deep collaboration as defined by DiMaggio and Powell. High levels of involvement in the collaborative process (i.e. the development of a strong mutual awareness that

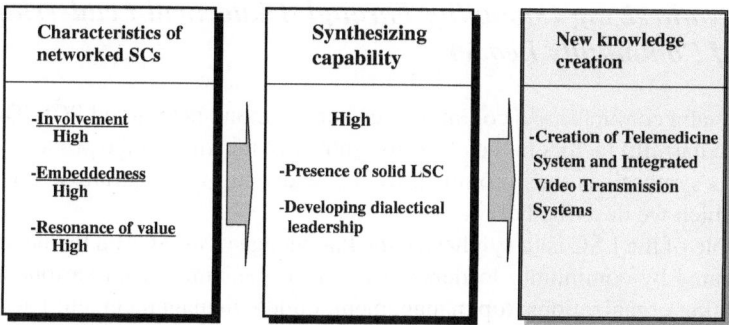

Fig. 5.5 Evaluations of new knowledge creation in the case

SC members are involved in a common enterprise) are also essential. Figure 5.5 shows the evaluation results of the networked SCs of 'Obihiro Project' with weight given to the factor of high involvement.

The second important elements in the formation of networked SCs – embeddedness-describes the degree to which a collaboration between SCs is enmeshed in interorganizational relationships. This element highlights the connection between the collaboration and the broader interorganizational network. Highly embedded collaborations between SCs were observed as an expanded network of SCs centered on how the Institute came into being.

The third important element in the formation of networked SCs is the resonance process of values in the community. This is the process whereby all community members, in their effort to fulfill the networked SCs mission and goals, share and resonate values aimed at achieving the business. This idea of the resonance of values is also the same as the hidden value, espoused by O'Reilley and Pfeffer, that enables the shared values within the formal organization or community to produce new knowledge or competencies(O'Reilly and Pfeffer 2000). The resonance of values in networked SCs with partners inside and outside the organization and in networked SCs with partners (kodama 2001) leads to dialectical ideas and strength to act among community leaders and turns into a capability for generating new knowledge that forms the new core for the networked SCs. High levels of resonance value were observed in the networked SCs of "Obihiro Project," which had succeeded in the large project (See Fig. 5.5).

For the analysis mentioned above, community members including community leaders need to build a platform for resonating values and creating relationships of mutual trust while also engaging in ongoing mutual exchanges, deep collaboration, high involvement and high embeddness at the boundaries of multiple, different SCs. An evaluation of networked SCs can be found in the representative pattern of the big project which has quickly networked their SCs and realized high involvement, high embeddness and high resonance value within these SCs. These effects can establish the leadership-based strategic community between community leaders and produce the synthesizing capability based on their dialectical leadership (See Fig. 5.5).

5.4.2 Synthesizing Capability Through Dialectical Leadership of Community Leaders

Struggles and conflicts are a common occurrence among networked SCs. These elements are harmful factors in the effort to synthesize the knowledge possessed by the SCs. This synthesis is thus promoted by the leadership-based strategic community (LSC) which we describe below.

The role of the LSC is to synthesize the knowledge of all SCs on the network that were formed by community leaders (made up of personnel from various levels of participating organizations: top management, middle management, etc.) and to generate synthesizing capability through dialectical leadership, the combined network power of all SCs (See Fig. 5.6). The LSC needs to balance the various paradoxical elements and issues within SCs on the network in order to realize this synthesizing capability. The LSC also needs to enable community leaders to consciously conduct dialectical management based on dialectical leadership and engage in dialectical dialogue to solve the various differences and issues that result from learning among the community leaders. As a result, the LSC actively analyzes problems and resolves issues, forms an arena for the resonance of new values, and creates a higher level of knowledge. Dialectical management is based on the Hegelian approach, which is a practical method of resolving conflict within an organization (Benson 1977; Peng and Nisbett 1999; Seo and Creed 2002).

Dialectic first appeared in the question and answer technique of Socrates and Plato's theory of ideas, and it became an approach to thinking about things that

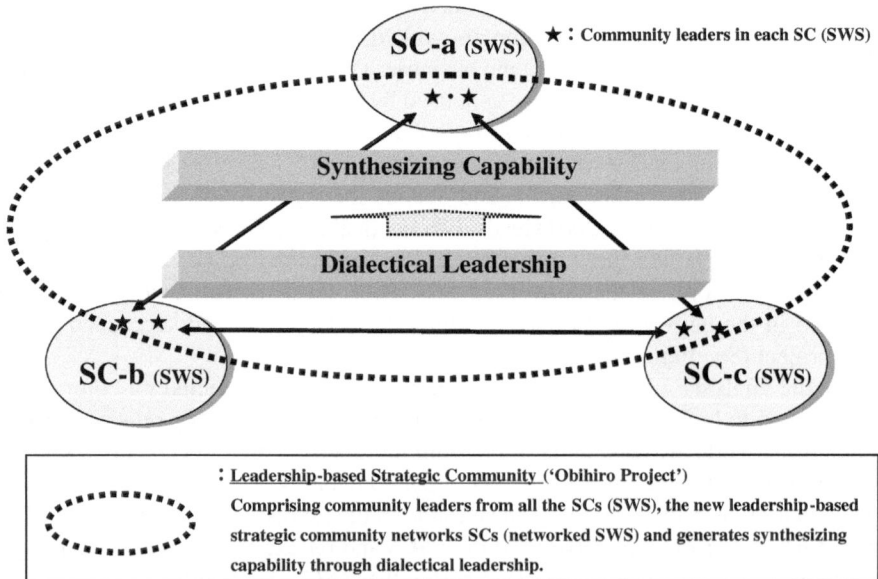

Fig. 5.6 Synthesizing capability through dialectical leadership of community leaders

was discussed and developed through the history of philosophy. In particular, Hegel (1927) considered dialectic to be a law of dynamic development in cognition and existence, proposing the thesis, antithesis, and synthesis scheme of logic and the concept of "aufheben" (to sublate). According to Marx (1930, 1967), Engels (1953), and others, Hegel's ideological dialectic developed in a practical methodology. They applied dialectic approaches to thinking to civilization and culture, produced thesis and antithesis with respect to propositions and historical fact, and proposed a methodology by which problem areas and conflicts were resolved through the synthesis of the two sides. The synthesis then became a new thesis, both of which were denied by antithesis, which produced another new thesis in a never-ending process; the process of historical development was proposed to be an eternal process. Dialectic, on the other hand, was also applied to organization theory, stimulating discussion based on absolute truths or morality in devotion to the community (Benson 1977) or in the process of corporate reform (Van de Ven and Poole 1995). In addition, Peng and Nisbett (1999) and Peng and Akutsu (2001) analyzed the psychological reactions that could easily result from two apparently contradictory propositions and, while risking crises that allow contradictions, proposed "dialectical thinking in a broad sense" that judged parts of both propositions to be correct. Recently Seo and Creed(2002) used a dialectical perspective to provide a unique framework for understanding institutional change that more fully captures its totalistic, historical, and dynamic nature, as well as fundamentally resolves a theoretical dilemma of institutional theory.

The balancing of paradoxical elements and issues involves the synthesis of mutually divergent views among organization members coming from different corporate cultures on the one hand, and the synthesis of a variety of divergent business as well as social issues (such as the procedures of different management, technologies, or customers' needs). In the case of "Obihiro Project", for example, synthesis were required in three areas: (1) the values of many employees possessing a broad diversity of viewpoints and knowledge shaped by the different organizational cultures to which they belong, (2) balancing customers' needs and high-technology presented by firms, (3) balancing political aspects and social needs. The LSC plays a central role in synthesizing the paradoxical elements and issues in the specific areas of human resources, elements among seeds and needs, political and social issues. The dialectical leadership with the new ideas and approaches of the community leaders who have adopted the methods of dialectical management based on dialectical leadership in their efforts to synthesize paradoxical elements and issues make new knowledge creation and innovation possible.

The LSC promotes dialectical dialogue and discussion among community leaders in order to cultivate a thorough understanding of problems and issues. By communicating and collaborating with each other, community leaders become aware of the roles and values of each other's work. As a result, community leaders are able to transform the various conflicts that have arisen among them into constructive conflicts (Robbins 1974). This process requires community leaders to follow a pattern of dialectical thought and action in which they ask themselves what sorts of actions they themselves would take, what sorts of strategies or tactics they would adopt, and what they could contribute toward achieving the large project and the innovation of a new knowledge creation. And in achieving new knowledge creation and innovation, the LSC promotes the sympathy and resonance of the community leaders' values, and the combined synergy and dialectical leadership among the community leaders have resulted in the high levels of synthesizing capability that has enabled "Obihiro Project" to realize this big project and create the Integrated Video Transmission System. In another sense, it can be seen that "Obihiro Project" has used the resonance of values among community leaders in their SCs and their leadership synergy based on dialectical leadership to form the LSC and high levels of synthesizing capability, which in turn generated a solid network of SCs. (see Fig. 5.5).

5.5 Managerial Implications: Toward the Realization of Strategic Community-Based Organizations

Through the analysis of an in-depth case study, I would like to emphasize that innovative organizations in the twenty-first century need to be strategic community-based organizations. In other words, I believe that it is important for innovative organizations to create ongoing innovation through business as well as non-profit activities that involve the formation of SCs based on creative knowledge assets and the networking of these SCs. Knowledge, or management resources, aimed at innovation is created from SCs, a wide range of knowledge both inside and outside the company, including customers and strategic partners, is synthesized via the network, and new knowledge that never existed before is created to become a new source of competitive advantage. To that end, it is important for community leaders who form the LSC to find a new value aimed at innovation with customers and leaders of strategic partners inside and outside the organization as the organization endeavors to achieve its desired vision and mission. The newly created value is then shared, sympathized, and resonated by all community members through dialectical dialogue and discussion within the LSC. The philosophy of an interactive learning-based strategic community, where members teach each other and learn from each other, is an important part of this process. This approach promotes further dialectical leadership based on dialectical consideration and becomes the driving force for producing high levels of synthesizing capability.

As a first important element regarding the dialectical leadership which synthesize the different leadership behaviors of leadership's dominant dualities, the community leaders in the LSC, who use synthesizing capability, do not only exhibit their "strategic leadership" as directors based on integrated, centralized leadership which can produce short&long-term strategy, focus on big picture and perform efficiency, but also exhibit their "creative leadership" based on autonomous, decentralized leadership which can produce creative thinking and behaviors of community members. As a second important element, community leaders do not only exhibit their "forceful leadership" as directors who can take charge and control community members, but also become listeners, recipients and collaborators based on collaborative leadership (Chrislip and Larson 1994; Bryson and Crosby 1992), empowering community members through enabling leadership and enhancing intrinsic motivation (Osterlof and Frey 2000) among community members in their knowledge creation activities. Their role as supporters and followers providing ongoing collaboration and support for the community so that it can pursue dreams and a sense of accomplishment for the business and its vision requires the element of "servant leadership" (Greanleaf 1979; Spears 1995). In this way, community members are themselves able to participate in decision-making in the SC and to enhance mutual understanding and strengthen links within the SC. As a result of synthesizing capability through dialectical leadership, community leaders including community members can create new knowledge creation of strategic, creative big project, business concepts, and reforms in business processes (See Fig. 5.7). This image of leadership at strategic community-based organizations is not the old type that was buttressed by a rigid hierarchy but a new model of leadership aimed at achieving innovation. This new type of leadership, that is the dialectical leadership, is oriented toward the growth of not only individuals but of groups or organizations in the form of SCs and networked SCs at the same time.

From the viewpoint of leadership and organizational development, the key issue for companies of the twenty-first century aiming to achieve innovation is how to nurture and produce many community leaders who possess the dialectical leadership based on the concepts of being creative and strategic, and the ability to act. And the

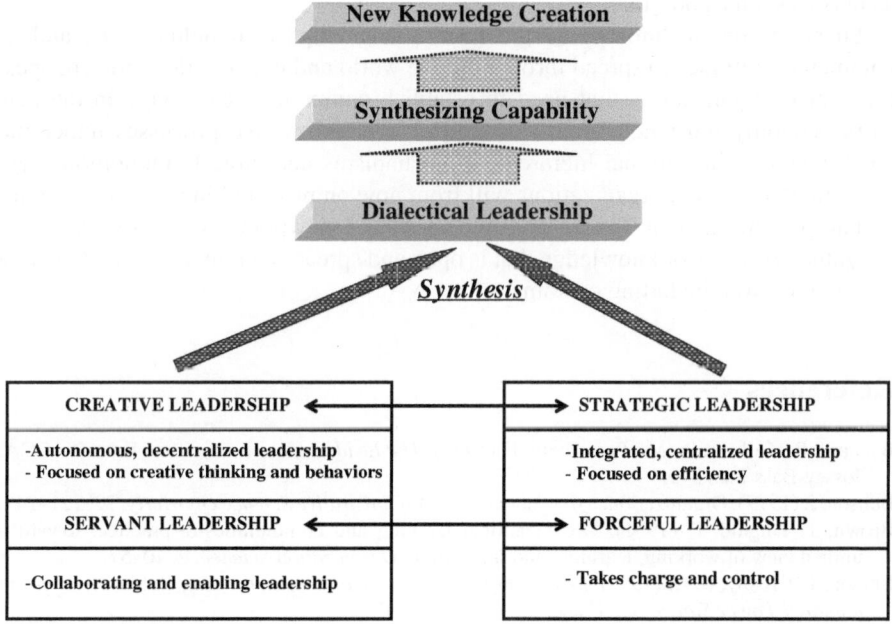

Fig. 5.7 New knowledge creation through dialectical leadership

formation of SCs that can continuously generate innovation via the capabilities of individual community leaders, the number of community leaders from each level of management, and the abilities of these leaders to exercise their skills, along with the networking of these SCs, determine whether or not a company is able to build synthesizing capabilities through dialectical leadership.

5.6 Conclusion

Through an in-depth case study, I presented one view on the proposition of what the capabilities of leading organizations in the knowledge-based society are for strategic community-based organizations that form dynamic innovative processes in SCs and network these SCs. In other words, one of the keys to producing innovation in a knowledge-based society is how organizations can organically and innovatively network different knowledge created from the formation of a variety of SCs inside and outside the organization, and acquire the synthesizing capability through dialectical leadership they need to generate new knowledge.

As community leaders, managers in the organization who play important roles in producing dialectical leadership for the organization use dialectical thinking and power to act to synthesize knowledge of good quality that was unevenly distributed inside and outside the organization. To this end, it is important for community leaders to promote the speedy formation of quality SCs networked inside and outside the organization, including customers, and to form an LSC made up of community leaders as soon as possible.

Superior core technology in the leading-edge high-tech fields of IT and e-commerce continues to spread throughout the world and undergo dramatic changes. Innovative organizations that need to establish competitive advantage in the network economy must not retain full control over innovative processes under the conditions of conventional hierarchical mechanisms and closed autonomous systems. In other words, organizations will from now on probably increasingly require a management that can, from a multiple variety of viewpoints, use networked SCs to synthesize superior knowledge that is open and spread out both inside and outside the organization, including customers.

References

Bechard, R., Goldsmith, M., Fesselbein, F. (1996). *The Leader of the Future*. San Francisco, CA: Jossey-Bass Inc., 3–9.

Benson, J. (1977). Organization: a dialectical view. *Administrative Science Quarterly*, 22, 221–242.

Brown, J., Duguid, P. (1991). Organizational learning and communities-of-practice: toward a unified view of working, learning, and innovation. *Organization Science*, 2, 40–57.

Bryson, J., Crosby, B. C. (1992). *Leadership for the Common Good: Tackling Public Problems in a Shared-Power World*. San Francisco, CA: Jossey-Bass.

Chrislip, D., Larson, C. (1994). *Collaborating Leadership: How Citizens and Civic Leaders Can Make a Difference*. San Francisco, CA: Jossey-Bass.

Dacin, M. T., Ventresca, M. J., Beal, B. D. (1999). The embeddedness of organizations: dialogure and directions. *Journal of Management*, 25, 317–356.

DiMaggio, P., Powell, W. (1983). The iron cage revisited: institutional isomorphism and collective rationality in institutional fields. *American Sociological Review*, 48, 147–160.

Engels, F. (1953). *Dialektik der Natur*. New York: International Publishers.

Granovetter, M. (1985). Economic action and social structure: the problem of embeddedness. *American Journal of Sociology*, 91, 481–510.

Grant, R. (1996). Prospering in dynamically competitive environments: organizational capability as knowledge integration. *Organization Science*, 7, 375–378.

Grant, R., Baden-Fuller, C. (1995). A Knowledge-based theory of inter-firm collaboration. *Academy of Management Best Paper Proceedings*, 38, 17–21.

Greanleaf, R. (1979). *Servant Leadership*. New York: Paulist Press.

Hegel, G. W. F. (1927). *System Der Philosophie Erster Teil Die Logik*.

Heide, J. (1994). Inter-organizational governance in marketing channels. *Journal of Marketing*, 50, 40–51.

Hesselbein, F., Goldsmith, M., Beckhard, R., Schbert, R. F. (1998). *The Community of the Future*. San Francisco, CA: Jossey-Bass Inc.

Hirose, T. (1985). Veterinary medicine – results and research prospects. Japan Veterinary Association, 233–237.

Kodama, M. (1999). Strategic business applications and new virtual knowledge-based business through community-based information network. *Information Management & Computer Security*, 7(4), 186–199.

Kodama, M. (2001). New business through strategic community management: case study of multimedia business field. *International Journal of Human Resource Management*, 11, 1062–1084.

Kodama, M. (2003). Transforming an old-economy company into a new economy- the case study of a mobile multimedia business in Japan. *Technovation*, 23(4), 239–250.

Lawrence, T., Phillips, N., Hardy, N. (1999). Watching whale-watching: a relational theory of organizational collaboration. *Journal of Applied Behavioral Science*, 35, 479–502.

Lipnack, J., Stamps, J. (1997). *Virtual Teams*. New York: John Wiley & Sons, Inc.

Marx, K. (1930). *Critique of Political Economy*. New York: Dutton.

Marx, K. (1967). *Writing of Young Marx on Philosophy and Society*. New York: Dutton.

Mintzberg, H., Ahlstrand, B., Lampel, J. (1998). *Strategy Safari: A Guided Tour Through the Wilds of Strategic Management*. New York: The Tree Press.

Nikkei Communication (1999). Potential of portable video phone. 298, 99–96.

Nippon Kogyo Shimbun (2000). Imaging protocol conversion system – NTT development. Feb. 4.

Nonaka, I., Konno, N. (1998). The concept of 'Ba': building a foundation for knowledge creation. *California Management Review*, 40, 40–54.

Nonaka, I., Toyama, R. (2002). A firm as a dialectical being: towards a dynamic theory of a firm. *Industrial and Corporate Change*, 11, 995–1009.

Ohira, H., Kodama, M., Yoshimoto, M. (2003). The development and impact on business of the world's first live video streaming distribution platform for 3G mobile videophone terminals. *International Journal of Electric Business*, 1(1), 94–105.

Osterlof, M., Frey, B. (2000). Motivation, knowledge transfer, and organizational forms. *Organization Science*, 11, 538–550.

O'Reilly, C., III, Pfeffer, J. (2000). *Hidden Value: How Great Companies Achieve Extraordinary Results with Ordinary People*. Boston, MA: Harvard Business School.

Peng, K., Akutsu, S. (2001). A mentality theory of knowledge creation and transfer: why some smart people resist new ideas and some don't. in *Managing Industrial Knowledge-Creation, Transfer and Utilization*. Nonaka, I., and Teece, D. (Eds),. London: SAGE Publications, 105–123.

Peng, K., Nisbett, R. (1999). Culture dialectics, and reasoning about contradiction. *American Psychologist*, 54, 741–754.

Porter, M. (1980). *Competitive Strategy: Techniques for Analyzing Industries and Competitors.* New York: Free Press.

Powell, W., Brantley, P. (1992). Competitive Cooperation in Biotechnology: Learning Through Networks?. in *Network and Organizations: Structure, Form and Action.* N. Noria and R. G. Eccles (Eds),. Boston, MA: Harvard Business School, 366–394.

Powell, W., Koput, K., Smith-Doerr, , L. (1996). Inter-organizational collaboration and the locus of innovation: networks of learning in biotechnology. *Administrative Science Quarterly*, 41, 116–146.

Robbins, S. (1974). *Managing Organizational Conflict: A Nontraditional Approach.* Englewood Cliffs, NJ: Prentice Hall.

Sawhney, M., Prandelli, E. (2000). Communities of creation: managing distributed innovation in turbulent markets. *California Management Review*, 42, 24–54.

Seo, M., Creed, W.D. (2002). Institutional contradictions, praxis, and institutional change: a dialectical perspective. *Academy of Management Review*, 27, 222–247.

Spears, L. (1995). *Reflections on Leadership.* New York: Wiley.

Storck, J., Patricia, A. (2000). Knowledge diffusion through strategic communities. *Sloan Management Review*, 41, 63–74.

Tokachi Mainichi News (1999a). Bidirectional data networks in veterinary medicine -construction of remote medical treatment system. Jan. 1.

Tokachi Mainichi News (1999c). Gigabit network practical research – project adopted from Obihiro. Dec. 4.

Van de Ven, A. H., Poole, M. S. (1995). Explaining development and change in organizations. *Academy of Management Review*, 20, 510–540.

Chapter 6
New Theoretical Framework and Insights Derived from Comparative Case Studies

6.1 New Business Creation by Transformation

6.1.1 The Importance of Change Management

Considering innovation and organizational change in big business to date, study groups centered on Tushman, Nadler, and Romanelli (Tushman et al. 1978, 1997, Tushman and Romanelli 1985; Nadler and Tushman 1989; Nadler et al. 1995; Romanelli and Tushman 1994) proposed a punctuated equilibrium model and mechanisms for innovation in many fields and business types. Brown and Eisenhardt (1998) describe strategy creation and implementation models that can handle competition, especially in leading edge fields, from the viewpoint of realization and continuity of change, that is, strategy for change designed to continually create a dominant position in a rapidly changing environment with many competitors.

One such model is Intel's practice of "time-pacing" (an adaptive strategy of scheduling before events occur). Time-pacing allows a company to compete according to scheduling at predictable, fixed intervals in rapidly changing markets where forecasting is difficult. Specifically, it is a routine-based, rhythmical strategy concept that anticipates events, and applies to the creation of new products and services at regular intervals. It can be surmised that time-pacing is responsible for the success of Intel and others in implementing simultaneous, agile reform of strategy, organization, technology, operation, and leadership at their own companies, and achieving timely and speedy congruence of these management elements. Many traditional companies and managers are more familiar with the strategy action mechanisms of event-based pacing (acting to support some kind of event) than of time-pacing. With event-based pacing, companies adjust their strategies in response to events such as the movement of competitors, technological change, deteriorating finances, or new client demands, and so implement reform and congruence of the elements of strategy, organization, technology, operation, and leadership.

As with the corporate innovation streams mentioned in Chap. 2, the content of organizational change is classified as either incremental or radical and discontinuous. Incremental change (the formulation and implementation of environment adaptive strategy) exhibits growth and increased efficiency of existing business with

regard to environment adaption, and radical or discontinuous change (the formulation and implementation of environment creation strategy) exhibits the effects of a
penchant for change or creativity. For companies to achieve such changes, the most
important issues are to simultaneously transform the elements of strategy, organization, technology, operation, and leadership while achieving congruence of these
management elements.

Looking to the future, securing sustained competitive dominance through the
synthesis of environment adaptive and creation strategies is key for companies in
high-tech businesses that can predict they will face multiple uncertain environments.
Because of this, it is becoming increasingly important to change the basic framework of a company's established strategy, organization, technology, operation, and
leadership. In order to bring about the discontinuous, major radical change indicated
by Tushman and Nadler, organizations undergo a process of change confronting
paradoxical situations among these five management elements in accordance with
their complexity and size (e.g., Kodama 2007a).

In this chapter, I will analyze issues that companies face in building new business
structures, including new products, services, and business models, from the viewpoint of the management elements of strategy, organization, technology, operation,
and leadership. Next, I will analyze the different contexts of each management element for companies that succeed with new business, and consider the key drivers
that enable congruence of the elements from the viewpoints of the "dynamic range
of knowledge boundaries" and the "process-based organization."

6.1.2 Issues in Promoting New Business: Non-congruence Among Individual Management Elements

I carried out in-depth interviews at 22 Japanese companies[1] with regard to the com-
plaint that "We cannot create a hit product, and our success rate in new business
is low. What are the reasons for this?" Based on this interview I analyzed in detail
considering the management elements of strategy, organization, technology, operation, and leadership. From the findings, I then collated core factors shared by each
company from a business context viewpoint (see Fig. 6.1)

To begin with the strategy and technology viewpoint, the results showed that
most respondents at every company commented that they had a strong technological
orientation but inadequate marketing concepts focusing on the customer. They also
revealed a failure of strategy through inconsistency between new strategy formulation and specific implementation, manifested in such ways as planning and drafting
new projects around new business start-ups and inadequate creation, incubation, and

---

[1]Interviews were conducted with 22 Japanese companies (12 manufacturing, 2 retail, and 8 IT).
The data collected related to problems and issues in-house, mainly with regard to new product and
business development activities. The interviews with senior executives and managers were carried
out based on conditions of anonymity.

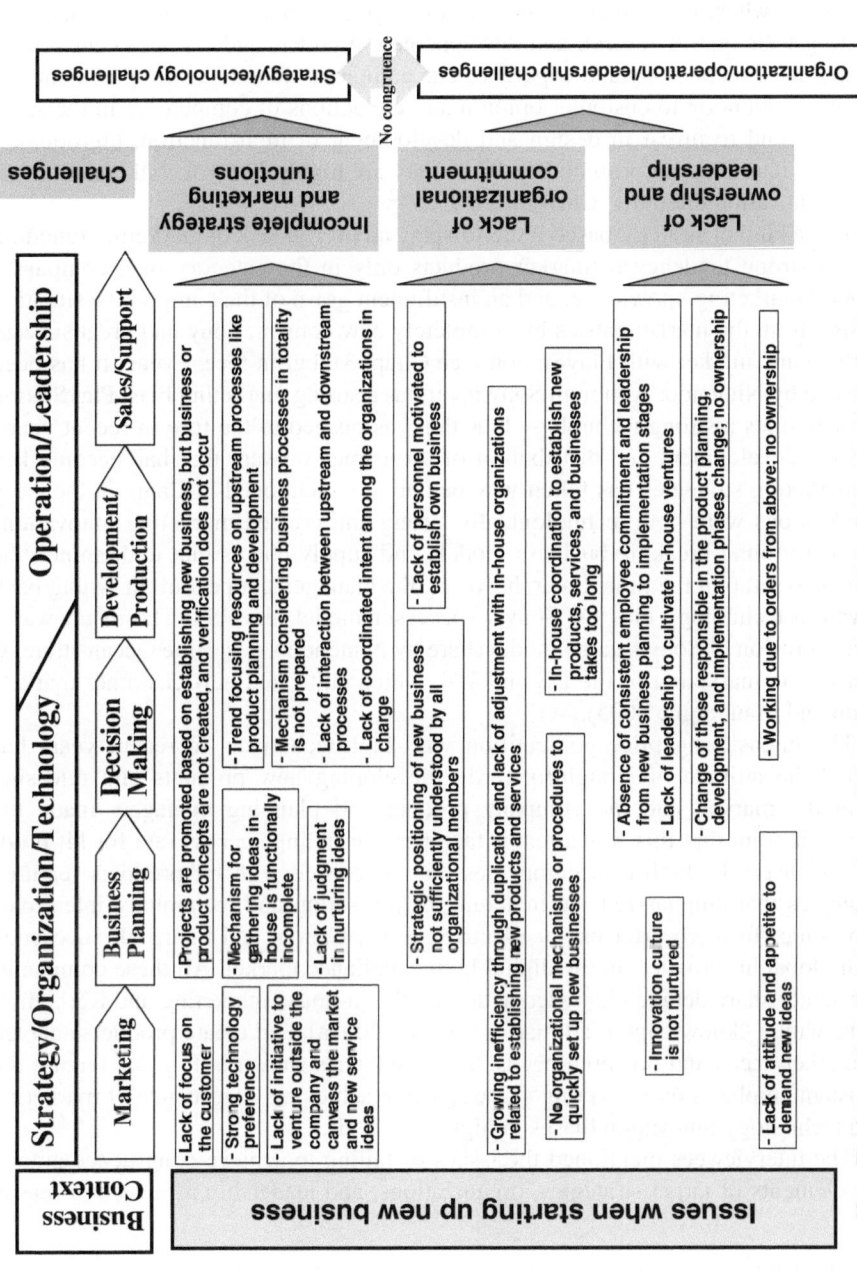

Fig. 6.1 Challenges driving new business (based on interview surveys)

Note: Key issues selected based on interviews at 22 Japanese companies (12 manufacturing, 8 IT, 2 retail)

inspection operations at the practice level for important business and product concepts. These tendencies are especially strong in the high-tech manufacturing and IT industries, where each company focuses on a specific marketing segment (specific, existing B2B or B2C users), and develops new products and services centered on accumulated technology and expertise. The companies feature a strong tendency to pay close attention to customer opinion and the actions of competitors in the same industry, and to invest in design and development of high-function, high-quality products that improve profitability. Thus they are highly likely to fall into the trap of an innovation dilemma (Christensen 1997)[2]

These kind of strategy-based issues display an overall lack of marketing function, with a strong tendency to market products only in the category of a company's general marketing knowledge, and an insufficient grasp of the competitive situation arising from the market entries by completely new comers. Sony had great success in the game market with PlayStation (see Chap. 3), but in recent years it has been pursued by Nintendo. Although Sony itself has a strong marketing bias, PlayStation 3 exemplifies a strong technology bias that has pushed to the fore in recent years, and the development and distribution of the game software that has become key to marketing strategies has fallen way behind. As mentioned in Chap. 3, the early PlayStations were both technologically strong and created marketing innovations through ground-breaking business models and supply chains that undermined the Nintendo setup of the time. Later, however, the management pendulum swung back toward a technology bias for Sony's successor model, the PS3. The result was a severe erosion of Sony's game model share by Nintendo's new Blue Ocean strategy, focused on marketing of the DS and Wii, which cultivated new customer markets (Kim and Mauborgne 2005).

The intensifying price competition with high-tech rivals in recent years has fueled the attitude of straightforwardly developing new products and releasing them the market, and the company product and planning managers made the common point that this was a major factor in the falling success rate for hit products. Conversely, high-tech companies that succeeded with hit products featured strategies that emphasized the matching of marketing and technology intentions. Panasonic (formerly Matsushita Electric), Sony, Canon, and Sharp all succeeded with global hit products in the digital home appliance market. All these companies constantly share detailed knowledge among the in-house marketing and R&D divisions with a "knowledge road map" (Kodama 2007a), and create products adapted to market needs and potential needs. The companies shown in Fig. 6.2 feature the constant emphasis on marketing strategy while achieving congruence of marketing and technology innovation biases in high-tech fields.

The interviewees mentioned the issues of failing to achieve congruence among the elements of target strategies, organizations, and leadership as a second reason

---

[2]The hard disk and excavating industries reveal examples of failures due to innovation dilemmas (Christensen 1997). The main causes for these failures include insufficient grasp of customer needs, insufficient consideration given to compatibility with organizational value standards, and lack of organizational initiatives for new business.

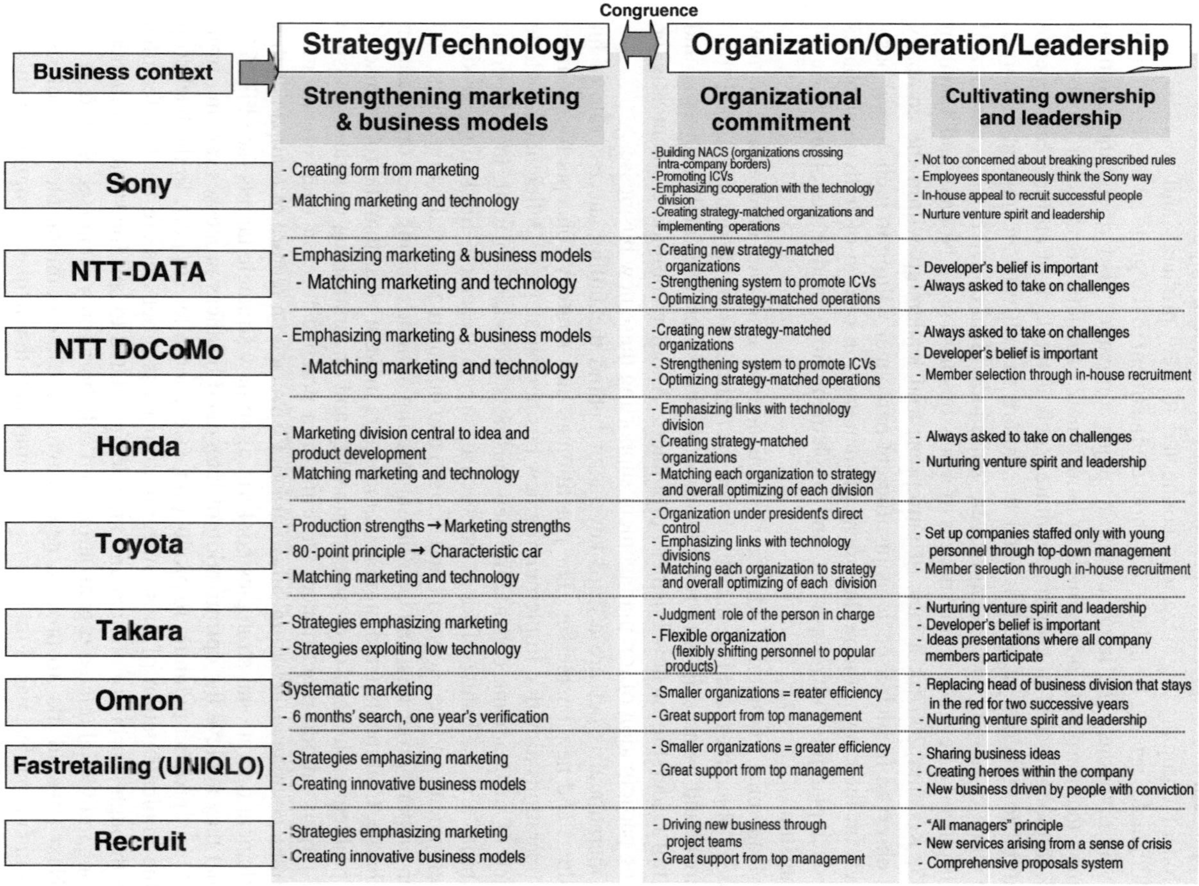
Congruence
Business context
Strategy/Technology
Organization/Operation/Leadership
Strengthening marketing & business models
Organizational commitment
Cultivating ownership and leadership
Sony
- Creating form from marketing
- Matching marketing and technology
-Building NACS (organizations crossing intra-company borders)
-Promoting ICVs
-Emphasizing cooperation with the technology division
-Creating strategy-matched organizations and implementing operations
- Not too concerned about breaking prescribed rules
- Employees spontaneously think the Sony way
- In-house appeal to recruit successful people
- Nurture venture spirit and leadership
NTT-DATA
- Emphasizing marketing & business models
- Matching marketing and technology
- Creating new strategy-matched organizations
- Strengthening system to promote ICVs
- Optimizing strategy-matched operations
- Developer's belief is important
- Always asked to take on challenges
NTT DoCoMo
- Emphasizing marketing & business models
-Matching marketing and technology
-Creating new strategy-matched organizations
- Strengthening system to promote ICVs
- Optimizing strategy-matched operations
- Always asked to take on challenges
- Developer's belief is important
- Member selection through in-house recruitment
Honda
- Marketing division central to idea and product development
- Matching marketing and technology
- Emphasizing links with technology division
- Creating strategy-matched organizations
- Matching each organization to strategy and overall optimizing of each division
- Always asked to take on challenges
- Nurturing venture spirit and leadership
Toyota
- Production strengths → Marketing strengths
- 80-point principle → Characteristic car
- Matching marketing and technology
- Organization under president's direct control
- Emphasizing links with technology divisions
- Matching each organization to strategy and overall optimizing of each division
- Set up companies staffed only with young personnel through top-down management
- Member selection through in-house recruitment
Takara
- Strategies emphasizing marketing
- Strategies exploiting low technology
- Judgment role of the person in charge
- Flexible organization (flexibly shifting personnel to popular products)
- Nurturing venture spirit and leadership
- Developer's belief is important
- Ideas presentations where all company members participate
Omron
Systematic marketing
- 6 months' search, one year's verification
- Smaller organizations = reater efficiency
- Great support from top management
- Replacing head of business division that stays in the red for two successive years
- Nurturing venture spirit and leadership
Fastretailing (UNIQLO)
- Strategies emphasizing marketing
- Creating innovative business models
- Smaller organizations = greater efficiency
- Great support from top management
- Sharing business ideas
- Creating heroes within the company
- New business driven by people with conviction
Recruit
- Strategies emphasizing marketing
- Creating innovative business models
- Driving new business through project teams
- Great support from top management
- "All managers" principle
- New services arising from a sense of crisis
- Comprehensive proposals system

Fig. 6.2 Business contexts of successful companies (based on interview surveys) behind the falling success rate of hit products and new businesses. Specific factors were, first, the issues of organizational commitment and company rules regarding new business. Standing out among these were a lack of organizational mechanisms and procedures for quickly establishing new businesses, duplication or cannibalization of company organizations connected to new products and services, and inefficiencies arising from lack of in-house adjustment to integrate awareness among relevant divisions. The interviewees at major companies also voiced the common opinion that corporate adjustment to launch new products, services, or businesses took too long. Issues of corporate culture, such as the lack of attitude and appetite required to generate new ideas, and the failure to cultivate an innovative corporate climate, were observed. Moreover, while individual convictions, feelings, and skills played a key role in the success of a new business (e.g., Nonaka and Takeuchi 1995; Kodama 2007b), a lack of entrepreneurial leadership within the company was observed in such aspects as insufficient personnel willing to launch new businesses on their own, the replacement of managers in charge of individual phases (product planning, development, and implementation), and a deficient sense of ownership.

Meanwhile, companies adopting organizational strategies conforming to new strategy objectives (see Fig. 6.2) displayed the features of ample organizational commitment and ownership with regard to new products, services, and businesses, and the full leadership of new business promoters. Another common element was the strong participation of top management supporting these new businesses.

Operational issues are a third key factor behind the falling success rate of hit products and new businesses. The practitioners lack strategy concepts for optimizing operations aimed at implementing new business strategies. Deficiencies include a lack of mechanisms considering new business processes in total, lack of compatibility and exchange of ideas between upstream (research, marketing, product development, and development and design) and downstream (production, sales, and after-support) processes, and insufficient meeting of minds among each organization in charge. Comments from numerous managers make it clear that the root cause of the problem is the lack of close commonality and understanding of strategic vision and targets among practitioners working on a practice level at each organization within a company (including subsidiaries). Conversely, a feature of companies that implement operational strategies conforming to new strategic targets (see Fig. 6.2) is that they achieve the optimization of integrated business processes ranging from R&D and product planning to production, sales, and after-support. In the backdrop to the implementation of these integrated business processes are the key features of a through sharing of context and knowledge related to specific strategy practices among individual related organizations and collaboration system (see Sect. 6.3).

Next I would like to provide new insights into the common elements of successful companies, based on analysis of companies that achieve congruence among the individual management elements of strategy, organization, technology, operation, and leadership, and succeed with new products, services, and business models (please refer to Fig. 6.2 and Columns I–V for leading companies in this area).

6.2 Comparative Case Analysis and New Insights

I obtained survey results noting incomplete strategy and marketing functions, inadequate organizational commitment, and inadequate ownership and leadership as issues driving the kind of new business shown in Fig. 6.1 (new products, services, and business models). These factors gave rise to non-congruence among the individual management elements of strategy, organization, technology, operation, and leadership. In this section, I would like to consider the success factors (see Fig. 6.2) of new business (new products, services, and business models), based on the management issues and challenges faced by these individual companies from the viewpoint of congruence of these five elements. With regard to linkage of these individual elements, which are central to the debate, numerous managers in each company indicated the importance of matching these elements to overall optimization (equivalent to the "boundaries congruence" mentioned in this book).

6.2.1 Congruence of Strategy and Technology

With regard to congruence of strategy and technology, managers at each company believe that synthesizing technology and market orientation oriented to new products and businesses is the most suitable strategic philosophy. In other words, managers recognize that formulating and implementing technology strategy through market-oriented thinking is essential. They expressed the following sentiments:

> Strategy is determined by synergies between market and technology orientations (Honda).
> You have to watch the market closely. It is especially important to consider the long- and short-term simultaneously, and exploit market-concept thinking (DoCoMo).
> It is important to consider essential technology elements and determine product strategies adapted to marketing (Canon).
> It is important to bounce marketing and ideas on technological innovations off each other (Panasonic).
> Marketing is the top priority, and it is important that the technology to realize this ties together globally outstanding companies and partnerships (SoftBank).

Meanwhile, numerous managers in the IT and information communications fields indicated that the diversification of business models of recent years had moved beyond the simple technological aspect of developing the product itself to bring to prominence the market (including the product) and business model strategies of the entire service. They commented as follows.

> Cultivating new markets isn't all about prioritizing technology. Original business models are the most important thing (DoCoMo).
> Creative services leading to breakthroughs require more than a technology angle. They need new frameworks concentrating knowledge from different industries (NTT-DATA).

Thus the key elements required for practitioners to achieve congruence of strategy and technology go beyond product and service development arising from the technological evolution of conventional science and engineering and the implementation of a strategy-making process aimed at business development to include wide-ranging technology (in the broad sense of the word, meaning procedures to realize new business) concepts of new business model ideas and market-oriented frameworks aimed at realizing targeted new business.

6.2.2 Congruence of Strategy and Organization

Regarding the congruence of strategy and organizations at each company, managers recognise that creating an optimal organization to implement appropriate strategies is the most important element in succeeding at new business. Specifically mentioned as key factors were positioning a new organization ("Can it be implemented within existing business divisions using existing resources; within the business division using the new organization; with the new organization directly controlled by the HQ; with a subsidiary; with a joint venture?") and its new organizational functions, scale and structure. A feature of the successful companies, as shown in Fig. 6.2, is the organizational commitment to realize new business strategies from the viewpoint of business model concept structures and market orientation reflecting the technology (in the broad sense of the word, meaning procedures to realize new business). The companies achieve this by building new organizational structures, such as dedicated projects, cross-functional teams, and internal corporate ventures (ICVs).

Furthermore, leaders and managers at every company believe that choice of specific personnel to promote business at the new organizations is important, and that key decision-making items in promoting new business involve such questions as "Who, with what skills, should we select and deploy—specialists, or additional posts at existing offices?" "Should we appoint an executive with the sole task of supervising the new organization, or make it an additional post for someone from the existing business HQ?" Corporate leaders and managers also indicated the importance of cultivating ownership and leadership of the people who are launching one's own new business or appointed, commenting that "Conviction among developers is important" (Takara Inc.), "It's important to leave new business initiatives to those who want to execute them" (Omron Inc.), and "New business arises from a sense of crisis" (Recruit Inc.).

Regarding other key items of strategic and organizational congruence, corporate managers strongly suggest the need, when implementing strategy, not just to select and deploy the optimal organizational forms and personnel but also to clarify business specializations at new and existing organizations while mutually linking individual organizations. Existing organizations imply management divisions (responsible for establishing strategy policies at HQ divisions and the planning divisions of business HQs) and business HQs (responsible for existing business duties), and the relationship between the existing and new organizations must be strengthened. Resource allocation including people, goods, money, information, and knowledge, and the mechanisms of optimizing the operations of business process implementation (mentioned below) are especially important to implementing new business within the company.

6.2.3 Congruence of Strategy and Operation

Numerous companies indicate the importance of exploiting the resource and routines of existing organizations to realize new business. Productive collaboration among new and existing organizations is an important factor in smoothly promoting new business, especially when implementing optimal operations at large corporations. The building of optimal business processes required to realize new business strategies by asking such questions as "To what extent should we have new organizations carry out operations?", "How should joint ventures maintain relationships with stakeholder companies?", and "How can we enhance synergies through collaboration with parent company?" is contributing greatly to realizing strategies more quickly and efficiently. Companies that are succeeding in new business (see Fig. 6.2) actively promote collaboration among the relevant responsible divisions aimed at optimizing the business processes to achieve strategies.

> We clarify the separation of duties among existing and new organizations within the company, and concentrate on carrying out the business responsibilities of individual divisions (DoCoMo).
>   We actively exploit assets as routines, including the operation and production functions of existing organizations (Sony, Honda, and Toyota). We methodically exploit the resources of parent company (Sony, Omron, DoCoMo, NTT-Data).

A feature of all these companies is that they achieve congruence of technology and organization aimed at realizing applicable strategies while adopting the operations best suited to these management elements (such as exploiting existing resources, building new supply chains, and implementing ICT management) and rebuilding the company's distinctive core competences.

6.2.4 Management and Leadership

The core practitioners in organizational leadership positions for establishing and implementing these strategy, organization, technology, and operation elements are top and middle managers at their respective companies. The common key elements among middle and top managers' thinking and actions in promoting congruence among individual management elements and succeeding with new business strategies indicated from interviews with CEOs, senior executives, and middle managers at the companies targeted in this survey are arranged in Table 6.1.

Considering these companies, I would like to describe important elements of management for corporate leaders and managers as the formation of SWS (small-world structures) and networked SWS gives rise to new business models and innovation. The issue of conditions required for top and middle management to achieve innovation concerns Japanese and foreign corporations (including multinationals) alike.

One key element for middle management is the need to understand and share context in the various SWS and networked SWS. The presence of common visions,

Table 6.1 Managerial implications on corporations

Important elements for middle management	Important elements for top management
1. Sharing and understanding diversity of context	1. Support for middle management
2. Improvisation	–Promote dialog with middle managers
3. Commitment	–Provide constructive support for middle managers
4. Shared value—Achieving the mission through resonance of value	2. Building a knowledge creation environment
5. Leadership	–Review personnel and remuneration systems
–Dialectical leadership	–Activate a rewards system
–Collaborative leadership	–Build an IT environment

interests, merits, and knowledge among the practitioners is essential. Another element is that of improvisation in the formation of SWS and networked SWS. Like musicians playing jazz or surfers riding waves, middle managers must make practical decisions instantly so as not to miss out on business opportunities (Kanter 2001). A third element is the need for middle managers in SWS and networked SWS to embrace a deep commitment as they aim to realize their vision. A fourth is value shared among practitioners as resonance (Kodama 2007b). A fifth concerns collaborative leadership among practitioners and balanced dialectical leadership centered on middle managers. Of course, these prerequisites are not easy for middle managers to commit to. But as they repeatedly engage in practice and self-examination, including organizational learning and challenge from failures, the required skills and know-how become deeply embedded in the practitioners as practical knowledge and experience.

Next, I will describe key top management elements. One is provision of support for the activities of middle managers in SWS and networked SWS. To this end, suitable executives in the top layer need to promote opportunities for deep dialog with middle managers while gaining a thorough understanding of middle managers' business activities and providing constructive support. A second element concerns the provision of a knowledge creation environment (e.g., Kodama 2008). Delivering an ICT environment, especially in global business, is essential for supporting the formation of SWS and networked SWS and for efficiently promoting business activities. In addition, top management should review personnel and remuneration systems and actively adopt reward systems as means of continuously maintaining the positive results of middle managers engaged in knowledge sharing, knowledge integration, and knowledge creation within corporations spanning formal organizations.

A key focal point common to successful companies is the implementation of collaborative and dialectical leadership through the formation of leader teams, cross-functional teams, projects and project networks, ICV organizations, and other SWS and multilayered, networked SWS within the company. Outside the company, too, these companies pursue dialectical leadership prioritizing the relationships among stakeholders inside and outside the company, as mentioned in Chaps. 3 and 4. Meanwhile, the formation of these SWS and networked SWS raises the recognition capability of practitioners around different contexts and knowledge areas; enhances the congruence of strategy, organization, technology, operation, and leadership; realizes optimal "business architecture" for corporate growth, and becomes a resource giving rise to corporate competitiveness (see the next section for details). Thus the formation of SWS and networked SWS is a key element for overall optimization through matching the individual management elements referred to by the leaders and managers at each company.

6.3   Crucial Elements for Boundaries Congruence and Successful Innovation

6.3.1   Creative/Productive Dialogue and Practitioners' Recognition Capability at SWS and Networked SWS

Innate thought-worlds (Dougherty 1992) and mental models (Spender 1990; Grinyer and McKiernan 1994) exist within each practitioner. Deeply embedded in the practitioners' minds and bodies are distinctive world views developed from past social experience, individual work functions as businesspeople, and practical and tacit knowledge gained from experience of each professional duty. The greater the degree of novelty and uncertainty of targeted strategic content, the more easily the mutual tension and conflict that arised from the diversity and difference of practitioners' mutual knowledge arises (Carlile 2002, 2004).

While transformation of practitioners' existing knowledge it is essential to the new business and product development innovations involving completely new concepts, as with the cases of the innovation dilemmas described in Sect. 6.1 above, the existing path-dependent knowledge deeply embedded in individual practitioners also has the potential to hinder the novelty of challenge. For new business development to succeed, practitioners forming SWS and networked SWS must first understand each other and generate new meaning by meaningful interaction in creative and productive dialog (see Fig. 6.3).

It is important for practitioners to deeply share and understand the newly created meaning involved in the novelty of new challenges through implementing deep dialog (e.g., Bohm 2004: Kodama 2007b). Communication and discussion among practitioners, however, is frequently peppered with exclamations of "I object!" and "I agree generally, but the devil is in the detail." Individual, specific discussions occur along the lines of "I understand what has to be done now, but specifically, who will do it, and in what way?" Tension and friction among organizations are inevitable in these kinds of discussions. However, the tension and friction with regard to current problems and the challenge of new issues also provide opportunities for innovation. Managers at each management level, including top management,

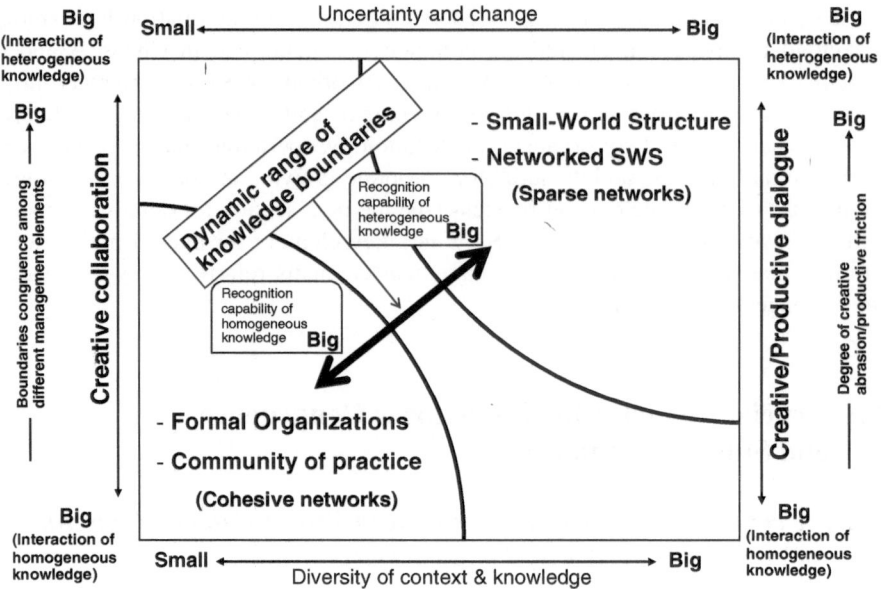

Fig. 6.3 The formation of knowledge boundaries and organizations

must actively and earnestly engage in deep dialog with practitioners as to where and why tension and friction exist, and what are the accompanying problems and means of solution. So as not to pluck the buds of new corporate growth, resolving tension and friction (including relations with external partners) through simple compromise is eschewed. Kunio Nakamura, chairman of Panasonic (formerly Matsushita Electric) states that we must not fear tension within the company in order to effect change (Kodama 2007c).

For tension and friction to become the driving force of corporate growth, practitioners must share common focal points, including establishing common interests (common merits of individual companies) (Kodama 2007c), building win-win relationships among stakeholders with partner incentives, sharing common strategic objectives and clarifying specific action plans. With regard to the in-house business context, it becomes important to clarify in-house decision-making processes and rules with regard to key items such as determining current and future priorities, action that needs to be taken now, and investments required for the future, and to disclose the results fairly to all employees. Unless this is achieved, trust and a sense of unity through creative collaboration among practitioners will not arise.

The formation of SWS and networked SWS establishes common ground (Bechky 2003) among practitioners, establishes mutual trust by transforming mutual friction and conflict into creative abrasion (Leonard-Barton 1995) and productive friction (Hagel and Brown 2005), and enables collaboration to be promoted at heterogeneous knowledge boundaries (see Fig. 6.3). Kanter (2001, p. 231) states that

"Conflict is seen as creative and something to be encouraged (instead of disruptive and something to be avoided)."

The creative and productive dialogs that form common interest and ground raise practitioners' recognition capability with regard to different contexts and knowledge while promoting practitioners' creative collaboration among different specialist fields and divisions (see Fig. 6.3). This in turn enhances common understanding and recognition capabilities with regard to the different contexts and knowledge among individual management elements (strategy, organization, technology, operation, and leadership), and promotes "boundaries congruence" among these elements (see Fig. 6.3). Promoting creative and productive dialog thus expands the practitioners' dynamic range, that is, the dynamic range of thought and action of the practitioners, with regard to changes in knowledge boundaries traversed by formal organizations, SWS, and networked SWS (referred to in this book as the "dynamic range of knowledge boundaries"), and enhances recognition capabilities for different areas of knowledge. Meanwhile the dynamic range of practitioners' knowledge boundaries enhances the recognition capability of homogeneous knowledge through sharing context and knowledge at formal organizations and communities of practice that implement daily tasks (see Fig. 6.3).

Surveys of successful companies (see Fig. 6.2) reveal the importance of achieving congruence of strategy and building new organizations to create innovation. One way of doing this involves building a new organizational culture that does not exist in the traditional organization. Mixed teams (including projects, project networks, CFTs, independent small-scale organizations, and ICVs) are formed from practitioners possessing diverse, heterogeneous capabilities and practitioners from different specializations and backgrounds. It is also important to assimilate personnel with different ideas from inside and outside the company (see, for example, the case of NTT DoCoMo in Chap. 4). Then these heterogeneous teams and organizations blow a fresh breeze to the existing traditional organizations, and impart new stimulation and inspiration (friction and conflict is also likely, of course) to a large number of practitioners aimed at breaking down the status quo, change, creation, and innovation. Innovation arises from mixed teams of SWS and networked SWS intersected by practitioners possessing heterogeneous cultures and specializations. At the intersection (Johansson 2004) of ideas from SWS and networked SWS, practitioners must boldly transcend their knowledge boundaries. The formation of SWS and networked SWS then expands the practitioners' peripheral vision (Day and Schoemaker 2005). Successful companies (see Fig. 6.2) constantly build the kind of environments that promote close collaboration between a new organization and the existing, traditional organizations through creative and productive dialog and provide continuous support from top management.

In sum, through in-depth case analysis and theoretical discussion, I have presented one view of the capabilities of leading companies in the knowledge-based society with regard to strategic organizations (the "process-based organization" mentioned in Sect. 6.3.2) that form dynamic innovative processes in SWS and networked SWS. One of the keys to producing innovation in a knowledge-based society is for companies to organically and innovatively network heterogeneous contexts and knowledge created from the formation of a variety of SWS and networked SWS inside and outside the company, and acquire new competences through innovative leadership by organizational management, including top & middle management.

In a business environment fraught with uncertainty and turbulent change, it is becoming increasingly important for corporate leaders and managers to create new knowledge for their goals through knowledge integration and transformation by dynamically forming (or rebuilding as needed) and networking SWS. I'd like to mention the corporate entities that employ dynamic processes promoting the continual, conscious formation of SWS & networked SWS among internal corporate organizations and partners, including customers, and become a driving force that generates incremental innovation and discontinuous (or radical) innovation as corporate innovation streams (described in Chap. 2).

6.3.2 Drawing in Knowledge Boundaries and the Practice Process at Process-Based Organizations

The "dynamic range of knowledge boundaries" (mentioned above) at once comprises the range of context and knowledge changes and that of recognition capability for the various values and diversity held by practitioners. At the companies shown in Fig. 6.2, all individual practitioners possess a broad dynamic range of knowledge boundaries. The extent of this range expands the dynamic range of practitioners' thought and action with regard to environmental change. Then creative and productive dialog broadens the dynamic range of thought for practitioners to recognize diverse contexts and knowledge with regard to SWS and networked SWS, and forms common ground and interest among practitioners. This broadened dynamic range of knowledge boundaries promotes the building of shared values and mutual trusts among practitioners (Kodama 2007b), commitment creation, and creative collaboration (Kodama 2007c). Next, creative collaboration solves urgent issues relating to organizations, technology, and operations oriented to realizing strategies aimed at new challenges, and achieves congruence among management elements. Thus creative collaboration through creative and productive dialog with regard to the organic organizational formations of SWS and networked SWS merges and integrates diverse contexts and knowledge, and promotes boundaries congruence by optimization among different management elements (see Fig. 6.3). The roots of this creative and productive dialog exist in the managers' collaborative and dialectical leadership (mentioned in Chap. 2).

Realizing corporate innovation streams that achieve transformation and congruence for the different contexts of the individual management elements (strategy, organization, technology, operation, and leadership) under a dynamically changing environment is a key corporate challenge (also mentioned in Chap. 2). It becomes essential for companies to formulate and implement integrative strategies that synthesize environment adaptive strategy (involving new business adapted to environmental change) with environment creation strategy (aimed at creating new environments from entirely new kinds of business). To achieve this, corporate leaders and managers need to improve their recognition capability for diverse context and knowledge relating to the management elements of strategy, organization, technology, operation, and leadership. Creating the ideal arrangement for the thoughts and actions of practitioners and optimized new organizations matched to this kind of complex strategy practice also become important.

Companies possessing practitioners with high recognition capabilities for different contexts and knowledge form "process-based organizations." These organizations are the images of organizations possessing the dynamic "practice process" of practitioners aiming to create new contexts and knowledge based on human network relationships in response to the creation and elimination processes of dynamically changing contexts. Thus the "process-based organization" can also be called the "context-based organization."

The basic form of the process-based organization's constituent elements arises from formal organizations comprising flat cohesive networks and SWS and networked SWS comprising sparse networks of practitioners from various formal organizations (see Fig. 6.4). SWS and networked SWS are also organizational bodies (network formations) that promote creative and productive dialog among practitioners at knowledge boundaries, enable practitioners to implement practice processes aimed at new challenges, and create new knowledge. Practitioners' specific tasks at the SWS and networked SWS focus on initiatives for integrating awareness within the company, creating ideas, and solving issues aimed at top management teams' decision-making on key strategy formulation and middle management teams' strategy formulation at a practice level. Deep debate and meeting of minds over key strategy formation takes place in connection with the business contexts that top and middle management teams target or have as interested in new innovation.[3] With task teams comprising general staff, moreover, debate and dialog on strategy formulation and business contexts relating to specific issues and problems take place at a more specific practice level to discover the most suitable direction and solution policies (see Fig. 6.4).

These kinds of SWS and networked SWS are network platforms with the purpose of expanding the dynamic range of practitioners' knowledge boundaries, enhancing the recognition capability of practitioners' heterogeneous contexts and knowledge, and dynamically combining and linking different knowledge boundaries aimed at new challenges and solutions.

The formal organization of the "process-based organization" does not imply an unchanging, stratified, hierarchical, and bureaucratic model (the image of a mechanical organization). The "process-based organization" possesses a flat structure (two to four layers) that establishes faster decision-making and transfer of authority (see Fig. 6.4). This flat formal organization comprises multiple business units. The specific tasks of practitioners at formal organizations are the key duties of reliably

---

[3] Closely integrated awareness and frank exchange of ideas among top and middle management teams are frequently seen among outstanding Japanese companies (e.g., Kodama 2007a).

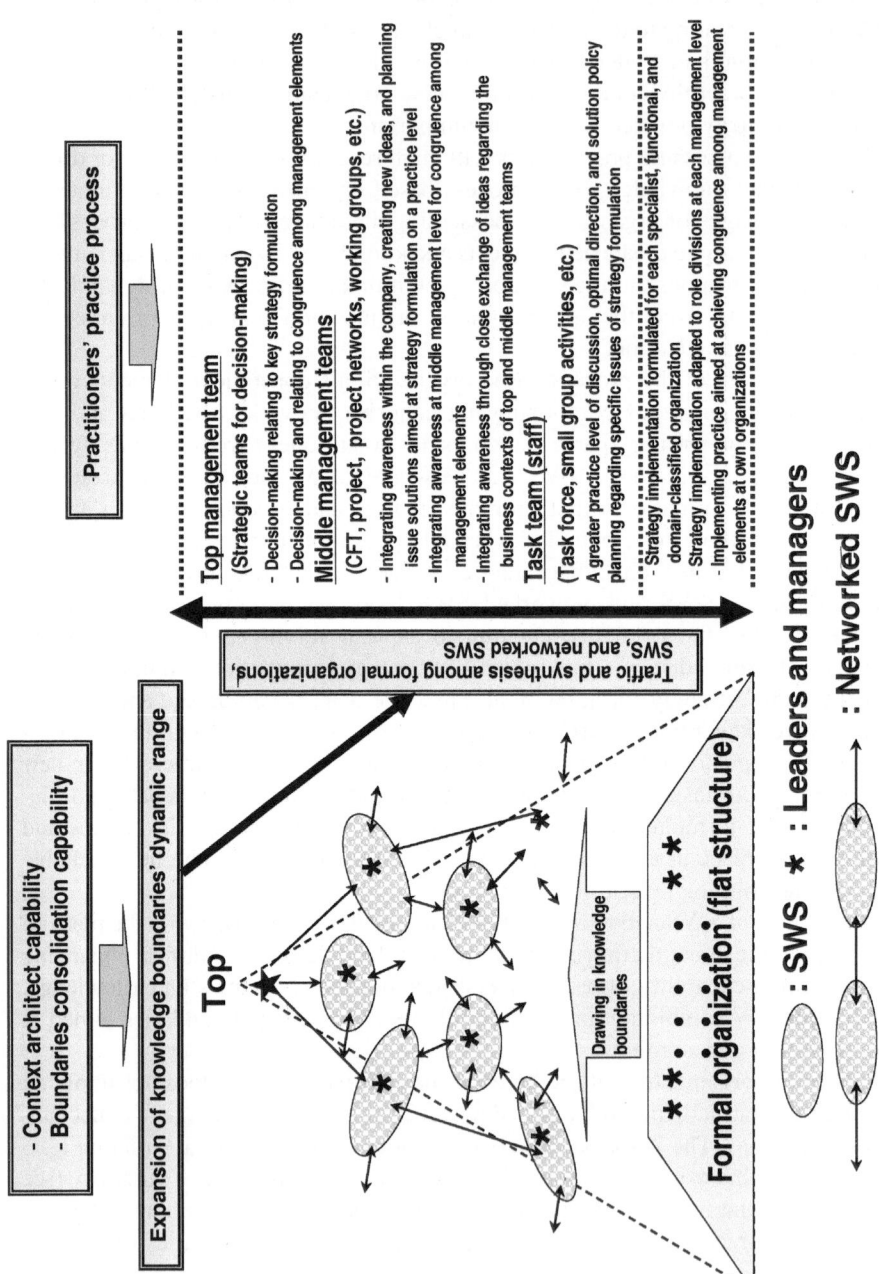

Fig. 6.4 Practice process for process-based organizations

implementing strategy practices of each business unit aimed at realizing strategies decided in practice processes at SWS and networked SWS. Moreover, communities of practice comprising cohesive networks with the same specialist area or work function, either within or crossing the boundaries of formal organizations, contribute to improving daily, routine tasks and operational efficiency. The leaders, managers, and staff of the business units collaborate with their counterparts in other business units in response to changing context while forming independently distributed SWS and networked SWS. SWS and networked SWS also embrace external partners and specific customers as required.

Outstanding leaders and managers expand the practitioners' own peripheral vision and create new knowledge through innovative leadership (Kodama 2007a) to form SWS and networked SWS and interact with customers, external partners and others as required. Knowledge is closely connected with situation, scene, and space; in other words, with context. In order to acquire, share, and create knowledge, practitioners must share context among one another. Moreover, specific contexts cannot be shared unless practitioners participate in specific SWS at specific time and places. Shared dynamic contexts are the medium that enables knowledge sharing and integration (e.g., Nonaka and Takeuchi 1995; Kodama 2007b). At such times, leaders and managers spontaneously form SWS as new knowledge boundaries, and draw their own stakeholder practitioners (the distinctive contexts and knowledge held by practitioners) into these SWS in order to link different contexts and create new contexts. I refer to this phenomenon as "drawing in to knowledge boundaries" (see Fig. 6.4).

This "drawing in to knowledge boundaries," is achieved through practitioners' "context architect capability" (Kodama 2009). To create new, high-quality knowledge, outstanding practitioners form specific SWS and draw in the practitioners that they require as stakeholders. The outstanding practitioners also spontaneously form or participate in multiple, heterogeneous SWS, circulate contexts and knowledge among the SWS, and take action to share and inspire context and knowledge among these practitioners. This is the practitioners' "boundaries consolidation capability" (Kodama 2009), whereby practitioners implement multiple, heterogeneous networked SWS. The "context architect" and "boundaries consolidation" capabilities create new concepts, ideas, and solution strategies required to formulate and implement targeting strategy, and go on to create dynamic practitioners' practice processes coming and going among formal organizations, SWS, and networked SWS (see Fig. 6.4).

6.3.3 Traffic and Synthesis Among Dual Networks

These SWS/networked SWS and formal organizations are not opposites. Rather, the process-based organization is the forum where the management of tasks within the range of formal-organization practitioners embraces SWS/networked SWS management methods that promote the sharing of different contexts and knowledge aimed at problem solving and realizing new business development (more advanced business such as creating ideas), and where the two management methods have a complementary relationship. The practitioners' practice process at process-based organizations emerges from the traffic among formal organizations and the SWS/networked SWS. Thus practitioners go on to implement targeted strategies while dynamically coming and going among the time-space intervals of context and knowledge, which transcend knowledge boundaries among the formal organizations and SWS/networked SWS.

Practitioners at process-based organizations not only comply with basic rules and decision-making processes for formal organizations and communities of practice within the company, but also manage challenging organizational behavior independently distributed at SWS and networked SWS while coordinating, expanding, and developing the SWS and networked SWS. The practitioners then adapt to the dynamic processes of changing context and knowledge, and come and go amid time and space intervals of context and knowledge transcending knowledge boundaries among the cohesive networks of formal organizations and the sparse networks of SWS and networked SWS. The "context architect" and "boundaries consolidation" capabilities possessed by the practitioners expand the dynamic range of the practitioners' knowledge boundaries, and promote the practitioners' "network connectivity action" among the different characters of the cohesive and sparse networks—the dual networks (see Fig. 6.5).

Considered from a business architecture point of view, traffic and synthesis among these dual networks has the following two implications (see Fig. 6.5). One is partial optimization among formal organizations at cohesive networks and individual management elements (strategy, organization, technology, operation, and leadership) at each formal organization connected to communities of practice. For congruence among management elements within the corporate system at SWS and networked SWS, practitioners promote optimization among the individual elements at their own organizations based on ample integration of awareness at related organizations and each management level.

The second implication is that practitioners decide to confirm the overall optimization (congruence and optimization among elements within the corporate system: see Fig. 6.1) of individual management elements transcending each formal organization by forming SWS and networked SWS. These are created based on the results of implementation to achieve congruence among each management element at formal organizations. The trafficking and syntheses of the dual networks by the practitioners merges and integrates diverse business contexts and knowledge inside and outside the organization, and goes on to build optimal business architecture as a corporate system.

Practitioners are liberated from the conventional management model thinking of formal organizations by considering organizational management from the viewpoint of SWS and networked SWS. A completely new management model then becomes important to the practitioners. This is a model whereby practitioners at formal organizations respond to dynamically changing contexts by coming and going amid the time-space intervals of context and knowledge; transcending knowledge boundaries among formal organizations and SWS/networked SWS; flexibly transforming the SWS and networked SWS in response to strategic objectives; and dynamically implementing strategy processes.

Moreover, the greater the complexity involved in new business and business model development and the difficulty of the concomitant problems and issues, the more SWS distribution and integration (networked SWS) becomes necessary. Thus practitioners integrate knowledge arising from different SWS distributed inside and outside the company by forming networked SWS. SWS are also organizational forms with objective-oriented and issue-solution type missions, and include the concept of organizational formation through project and cross-functional teams (CFTs). The organizational behavior whereby multiple projects build project networks (Kodama 2007c), and different project members collaborate to build new business models also corresponds to these networked SWS. The "process-based organization" enhances the extent to which practitioners form SWS and networked SWS and come and go amid the time-space intervals of context and knowledge transcending knowledge boundaries. The "process-based organization" then enhances the practitioners' behavior to dynamically create new knowledge.

Looking at process-based organizations on a time-axis of strategy practice, practitioners implement tasks (such as improved or reformed business procedures) for strategy implementation within their own range at the cohesive networks of formal organizations while undertaking difficult problem-solving tasks, and high-quality, demanding business challenges such as developing new products, services, and

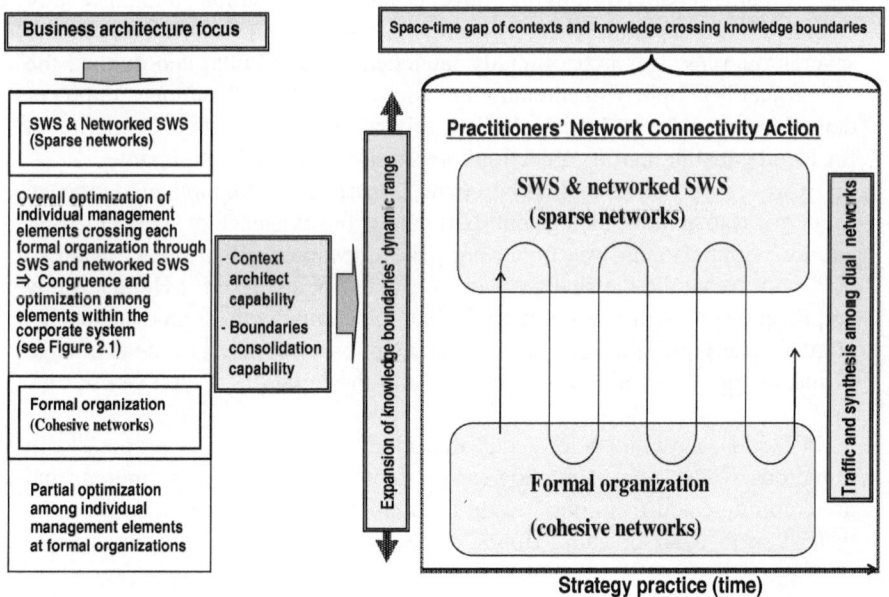

Fig. 6.5 Practitioners' network connectivity action

businesses at sparse networks of SWS and networked SWS (see Fig. 6.5). Then the process-based organization builds corporate innovation streams (see Fig. 2.4) to move forward with innovations that systematically enhance the efficiency of existing business, with the objective of creating short-term profit together with growth and long-term business. The strategic objectives of the process-based organization are also the optimal organizational formation for adapting to dynamic environmental change while creating one's own new environments and developing new business.

Column I Honda

Features of Honda's new product development originate from a "marketing orientation that systematically exploits ideas." This might include gathering every piece of marketing data, sharing market perspectives among people with different values (those in other positions), and debating the meanings among themselves centered on the marketing division in the HQ sales development department. The information is always dispatched to the management layer. Previously there had been examples of marketing products on the strength of technological novelty alone, leading to failure, but now marketing block personnel centered on outstanding, motivated social science and humanity graduates up to their early 30s have become the fulcrum of operations.

Product development revolves around marketing in a matrix organization. A motorbike product development project for young people, the "N project," is a specific example. Enthusiastic members were recruited within the company, separated from the conventional production line, and physically "isolated" in a separate place. The project, which comprised 20 youths and several veterans, was first officially launched in May 2005, and dubbed the "N project." People with dreams and ideas grapple with unique, unprecedented ideas on the periphery. At first, this was also a "heretical" experiment for Honda, and numerous objections were raised within the company.

However, the parties in this "heretical" experiment were motivated to make every effort to achieve their intent. Ultimately, the existence of such a tolerant culture within Honda was important. The questions "What do we want to do?" and "What did we change?" are constantly being asked at Honda. Many employees feel, with a sense of crisis, that the company must innovate or die. Honda's energetic spirit and sense of challenge is embodied in the challenge of the slogan "Do you have a Honda?" and the reply "I see you've got one, too!"

In Honda, as with "N project," this kind of in-house trial progresses in numerous ways. New technologies, products, and techniques emerge, and are communicated from one person to another. As ripples spread out on the water's surface, so the entire Honda company is undergoing constant change.

Column II Toyota

The environment surrounding Toyota was one of economic slowdown, increasingly borderless market economies, and the rapid development of information technologies. Toyota's chairman Hiroshi Okuda said the following:

> Companies must constantly reinvent themselves. Countless companies have collapsed because they were unable to free themselves from the successful patterns of the past. Even if you succeed, you must reform to become still better. Put another way, the organization that doesn't change will decay.

Because of the persistent long-term trend of young people shunning the company, in 1997 Toyota created the VVC (Virtual Venture Company). The 35 personnel centered on young people within the company and external personnel with different functions. The president suggested that the company spend three months considering measures that would appeal to the young. On this basis, Toyota implemented ground-breaking internal recruitment and selected 35 young personnel. Okuda said, "We leave cars for the young to the young." So that they could work unimpeded, he separated the group from the business division organization, and undertook to deal with them directly. VVC's in-house presentations featured hip-hop background music and Aloha shirts as the development team competed with ad agencies. Development time for new cars was reduced to 12 months, half that of conventional teams, and the team carried out its own, youth-oriented user research.

Toyota's VVC experience teaches three important lessons. One is that while rapid change is painful, it is essential to think on one's feet; second, that rough-and-ready may be preferable to highly detailed (one action beat a hundred discussions); and third, freely competitive power is true competitive power. The bigger the business, the greater the number of rivals—this truly denotes the pinnacle of business. It is important to constantly create competitive conditions within one's own company. Toyota's creation of VVC stimulated other Toyota organizations to create the Vitz and Altessa.

The merged brand "WiLL" was also founded on a proposal from Toyota. It was launched in August 1999 with the participation of Asahi Beer, Kao, KNT, and Matsushita Electric Industrial (now Panasonic). Kokuyo and Ezaki Glico jumped on board the following year, and WiLL developed as a project merging heterogeneous industries. With this WiLL joint venture, VVC investigated marketing techniques outside the automobile industry, and began to create new markets amid advancing information exchange from companies in different industries. WiLL arose from a sense of crisis over young people's alienation, and started up as a project within Toyota seeking out new marketing techniques. The participating companies also came to grips with the objectives of acquiring marketing techniques for the youth segment (which was not their specialty) and improving brand image. While this trial did not produce innovative success stories, it was highly stimulating and created new meaning embodied in the spirit of challenge, not just for Toyota's corporate culture but also for participating companies.

Column III Omron

Omron is a major Japanese corporation focused on the manufacture of control equipment and electronic components. Since its founding, Omron has considered what is required for the convenience and comfort of people and society, and has developed diverse businesses in accordance with the aims of its corporate charter created by founder Kazuma Tateishi: "To improve our daily lives through our own efforts, and create a better society." The "human viewpoint" of individuals residing in the community and "social needs creation" generating new values is rooted in Omron's corporate DNA.

At Omron it has been suggested that small-scale organizations create efficiency. One senior executive tells a story: "Three people carry stones to build a castle. One aims to make money, another to build something beautiful, a third to protect his country. They are doing the same work, but the outcomes will be completely different. Making the organization small confers responsibility. Even the workers who clean the office will suddenly change when a company becomes a subsidiary and that person is elevated to the post of manager or executive. But team design is always important. Quality, not quantity, is key when building a business. Appointing management, technology, planning and other staff and encouraging teamwork is essential. Until the team is established, leave it completely up to them."

Omron's basic criterion when judging whether to undertake new business is thought to be a yield rate of around 10 percent. Generally, investigation takes about six months, verification one year, and start-up 18 months. If two consecutive years of losses occur once the business is launched, the senior executive manager might be replaced. When launching a business selling photographic sticker devices, for example, Omron first advanced the technology of its face recognition system to produce a "print club" stickers device for photographic portrait. The device completely failed to sell. Omron had employed a survey company to canvass female junior high- and high-school students, but it hadn't done a good job.

Omron hired digital art specialists to create various pictures, and then conducted a survey of impressions, which indicated that very few items got high points. So as with items in a daily newspaper, Omron constantly replaced unpopular items with new. Sales fell for the industry as a whole, but Omron's sales doubled, achieving a market share of 52 percent within around five years of startup. Market surveys and rapid adaptation had been essential to Omron's success. The marketing of the original face recognition system leading to later digitization (previously this was an analog industry) of the sticker device happily answered many of the market's needs.

Omron's positioning of itself as a perpetual venture company emphasizes persistent challenge. Omron is convinced that displaying a challenging spirit to buck the status quo and pursue ground-breaking results in all areas of corporate activity without fear of failure is essential for the company to contribute to society.

Column IV Takara

Takara (which merged with Tomy in 2005 to become Takara Tomy) is a major Japanese planning, manufacturing, and marketing company whose products include toys, miscellaneous goods, card games, home games software, and infant-related goods. Takara proves that it is the vehicle to expand a range of exciting experiences, including toys, children's play and creative experiences. Many of its exciting experiences have a huge influence on children's growth, and consider children's sound development. Since its founding, Takara has consistently taken pride in its strong sense of mission in working for the children's future. It has developed high-quality toys throughout the world that embody the core concept of a "toy renaissance" in order to regenerate and develop the toy industry, and enhanced education, culture, and science while contributing to peace.

With Takara's new product development, the final judgment role within the company is taken by the head of the business division. Concerning the "Beyblade" product, Beyblade's success was linked to the conviction of a single person responsible at the business division. This person spent a year traveling to and from toy shops in his own time, interacting with children as would a store assistant, and pushed forward with the Beyblade development project with the conviction that it would be a sure-fire hit. The employee's belief in the product was a key factor in its success. The employees' experience of success itself is also important, and Takara believes that having such experience is vital to nurture talent. One example is the selection of a 39-year-old employee as president of Takara USA following success in Japan with "e-kara," a karaoke device with integrated microphone.

With a lineup of 1,500 products, the division between those that sell and those that don't is clear. Takara has adopted a reversible organizational system where personnel resources are used flexibly within the company, whether attached to the business or sales department. This organizational system has the benefit of being able to shift personnel flexibly to saleable products. Presentations of ideas are given twice a year to raise employee motivation. Each presentation offers 400–500 cases from a similar number of employees, and after judgments have been made, 50–60 people give presentations all day (each one lasting from 1 to 5 min) to the president and those responsible in the business division. The presentations have a fun, congenial atmosphere where presenters might use costumes or other items they have made themselves. Takara continues to challenge its boundless market and cultivate successive new markets from the new business models created by its distinctive corporate culture and venture spirit. Examples include expanding peripheral businesses in line with children's lifestyles, product development aimed at a wide age range, and further global expansion.

Column V Fast Retailing

Fast Retailing controls a range of companies including the clothing company UNIQLO. Founder, Chairman and CEO Tadashi Yanai was behind the move to target clothing companies with a global range, such as GAP. Now the group is expanding thanks to aggressive overseas development and M&As. Fast Retailing has set itself the following mission and vision:

- To continuously offer fashionable, high-quality, basic-casual clothing to be worn anywhere, anytime, by anyone at the lowest possible market price.
- To achieve this, we will sustain low-cost management, maintaining the shortest and cheapest route between production and sales.
- To thoroughly consider customer services required of the company, and realize the best services.
- To provide an environment that delights people the world over, and perform innovative work in a non-bureaucratic, companionable team. We aim for results of high sales and

At Fast Retailing, the president encourages new products and business initiatives by, for example, supporting people who want to branch out into completely different service areas (such as selling vegetables). To keep the company's lively innovative spirit alive and anchored to its founding principles, employees recall the 23 management provisions when making key decisions. In order to enable correct judgments, Mr. Yanai verifies one provision each week to management, and shares how he, as founder, created the management concept.

In the past, a failure to improve results would trigger reform within the company. The slogan for in-house reform was "ABC," or "All Better Change" (UNIQLO even offered one million yen to customers in order to gather complaints). Such reform is the most important element in a company's rapid growth, and to this end the company has reviewed its businesses comprehensively, down to the most common-sense matters, and created a number of investigative projects in-house. With new product development, moreover, weekly meetings take place to share information on product development work, where the employees concerned sit around a table and share information. The company also emphasizes the flow model of knowledge management exploiting IT to the full. Exploiting individual diversity and achieving good results through teamwork are features of Fast Retailing.

Aiming to promote information sharing and motivation among all employees, Fast Retailing aggressively transmits information from executive management. For example, the company holds twice-yearly conventions of all store managers, and monthly pep talks (head office supervisors). It distributes videos to all stores, creates in-house bulletins (once every three months) to share success stories, and promotes the creation of "heroes." Creating heroes through awards, and sharing the reasons for the awards with all employees has become important. It is easy for other employees to understand iconic employee as heroes, and easy to spread and promote to the stores. Choosing the correct heroes is also important.

In this way, Fast Retailing's distinctive corporate culture and activities have led to the creation of truly good clothing with unprecedented value. The company has also contributed to people's richer daily lives and corporate development in harmony with the community.

References

Bechky, B. (2003). Sharing meaning across occupational communities: the transformation of understanding on a production. *Organization Science*, 14(3), 312–330.

Bohm, D. (2004). *On Dialogue*. 2 ed., London: Routledge.

Brown, S. L., Eisenhardt, K. M. (1998). *Competing on the Edge*. Boston, MA: Harvard Business School.

Carlile, P. (2002). A pragmatic view of knowledge and boundaries: boundary objects in new product development. *Organization Science*, 13(4), 442–455.

Carlile, P. (2004). Transferring, translating, and transforming: an integrative framework for managing knowledge across boundaries. *Organization Science*, 15(5), 555–568.

Christensen, C. M. (1997). *The Innovator's Dilemma: When New Technologies Cause Great Firms to Fail.* Boston, MA: Harvard Business School Press.

Day, G., Schoemaker, P. J. (2005). Scanning the periphery. *Harvard Business Review*, 83(11), 135–148, November.

Dougherty, D. (1992). Interpretive barriers to successful product innovation in large firms. *Organization Science*, 3(2), 179–202.

Grinyer, P., McKiernan, P. (1994). Triggering major and sustained changes in stagnating companies', in strategic groups. *Strategic Moves and Performance.* Daems, H. and Thomas, H., (eds.). New York: Pergamon, 173–195.

Hagel, J., III, Brown, J. S. (2005). Productive friction. *Harvard Business Review*, 83(2), 139–145.

Johansson, F. (2004). *The Medici Effect.* Boston, MA: Harvard Business School Press.

Kanter, M. R. (2001). *Evolve! Succeeding in the Digital Culture of Tomorrow.* Boston, MA: Harvard Business School Press.

Kim, W. C., Mauborgne, R. (2005). *Blue Ocean Strategy.* Boston, MA: Harvard Business School Publishing.

Kodama, M. (2007a). *The Strategic Community-Based Firm.* London: Palgrave Macmillan.

Kodama, M. (2007b). *Knowledge Innovation – Strategic Management as Practice.* Cheltenham: Edward Elgar Publishing.

Kodama, M. (2007c). *Project-Based Organization in the Knowledge-Based Society.* London: Imperial College Press.

Kodama, M. (2008). *New Knowledge Creation through ICT Dynamic Capability-Creating Knowledge Communities Using Broadband.* Charlotte, NC: Information Age Publishing.

Kodama, M. (2009). *Innovation Networks in Knowledge-Based Firm – Developing ICT-Based Integrative Competences.* Cheltenham: Edward Elgar Publishing.

Leonard-Barton, D. (1995). *Wellsprings of Knowledge: Building and Sustaining the Sources of Innovation.* Boston, MA: Harvard Business School Press.

Nadler, D. A., Shaw, R. B., Walton, A. E. (1995). *Discontinuous Change: Leading Organizational Transformation.* San Francisco, CA: Jossey-Bass.

Nadler, D. A., Tushman, M. L. (1989). Organizational framebending: principles for managing reorientation. *Academy of Management Executives*, 3(3), 194–202.

Nonaka, I., Takeuchi, H. (1995). *The Knowledge-Creating Company.* New York: Oxford University Press.

Romanelli, E., Tushman, M. L. (1994). Organizational transformation as punctuated equilibrium: an empirical test. *Academy of Management Journal*, 3, 1141–1166.

Spender, C. (1990). *Industry Recipes: An Enquiry into the Nature and Sources of Managerial Judgement.* Oxford: Basil Blackwell.

Tushman, M., Nadler, D. (1978). Information processing as a integrating concept in organizational design. *Academy of Management Review*, 3(3), 613–624.

Tushman, M. L., O'Reilly, C. A. (1997). *Winning Through Innovation* Cambridge, MA: Harvard Business School Press.

Tushman, M., Romanelli, E. (1985). Organizatinal evolution: a metamorphosis model of convergence and reorientation. *Research in Organizational Behavior*, 7(2), 171–222.

Chapter 7
Theoretical and Managerial Implications

7.1 Boundaries Management Frameworks

First I would like to consider boundaries management frameworks within the framework of the dynamic view of the strategic management process (see Fig. 1.1) through the case analyses presented in Chaps. 3–6.

As explained previously, the concept of "boundaries congruence" inside and outside the corporate system is important to achieving dynamic strategic management, while the presence of optimal "business architecture" with regard to environmental change and management elements within the corporate system (including strategy, organization, technology, operation, and leadership) is essential. In corporate innovation streams, key elements were the creation of new business models from environment creation strategy, and reform and growth of existing business adapted to environmental change through environment adaptive strategy. How companies achieve their corporate strategy aims through the practice of these corporate innovation streams is a big issue.

The question of how companies consider congruence with the environment, and how they dynamically change corporate boundaries and adapt to the environment (or create new environments) is an important theme for the implementation aspects of corporate strategy. Because of this, companies must have a corporate vision and strategic aim supporting environment adaption and creation (see Fig. 7.1). Outstanding companies share values and mission statements based on corporate visions deeply and continuously with all employees (Peters and Waterman 1982; Collins and Porras 1994; O'Reilly and Pfeffer 2000). It is important that firm strategic intent (Hamel and Prahalad 1989) around a core of top management ensures that short- and long-term business road maps (which may be knowledge road maps) (Kodama 2007a) are clearly drawn, and that meaning and significance oriented to achieving strategic aims that specifically break down this content are deeply shared among all employees. Practitioners' deep sense making (Weick 1995) and boundaries vision with regard to environmental change promotes the establishment of these corporate visions and strategic aims.

M. Kodama, *Boundary Management*, DOI 10.1007/978-3-642-03789-4_7,
© Springer-Verlag Berlin Heidelberg 2010

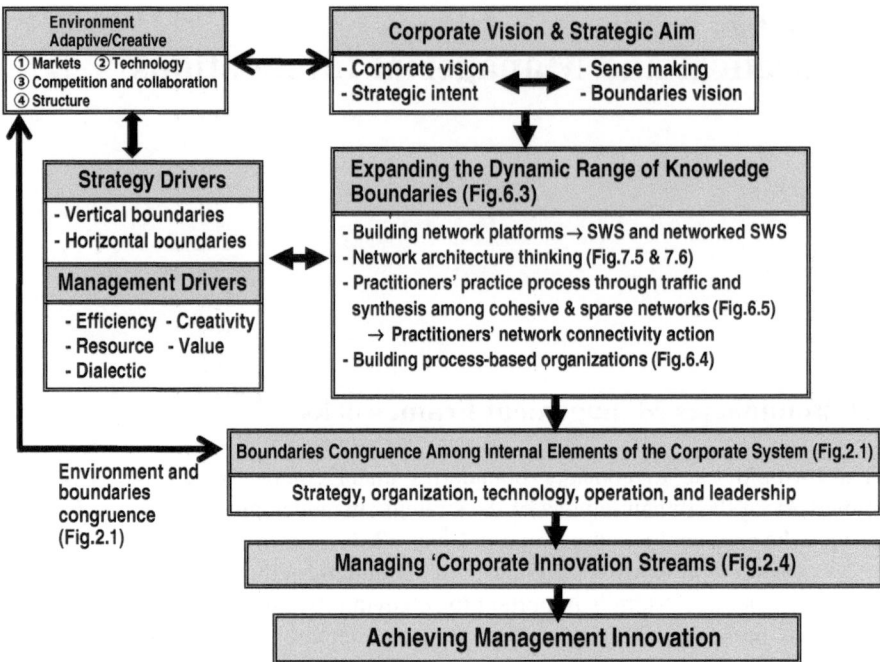

Fig. 7.1 Boundaries management frameworks

"Boundaries vision" means not only the peripheral vision of a company's core business domains, but also developing practitioners' vision towards unrelated industrial domains widely separated from one's own industry. This boundaries vision is significant because in recent years it has spurred the creation of new industry-transcending business models and new products merging different technology fields, and spurred the realization of environment creation strategy.

To establish this kind of corporate vision and strategic aim, companies must determine the strategy aims of sustainably competitive product, services and business models, and implement optimal design of vertical boundaries (value chains to achieve company-determined strategy aims) and horizontal boundaries (expansion and diversification of business domains) to achieve them.

To build an optimal value chain, a company must determine the vertical and horizontal boundaries of its corporate strategy, referred to as "strategy drivers" in this book (e.g., Santos and Eisenhardt 2005). One such driver is an industry value chain to realize the strategic aims of corporate strategy (Porter 1985). This strategy driver comprises business activities to form industry value chains, and is also a factor determining a company's vertical boundaries. A second strategy driver comprises the factors that determine a company's horizontal boundaries: the expansion and diversification (or reduction through selecting and focusing) of a company's existing business domains (products, services, and business models), or the creation of new business domains through merging different technologies. Companies must determine the strategic objectives of sustainably competitive product, service, and business models, and implement optimal design (architecture) of the vertical and horizontal boundaries as the strategy driver elements to realize these objectives.

The investigative elements that become important amid the corporate strategy formulation processes are the management drivers (Kodama 2009).[1] Companies have to consider, especially, optimization of the individual element parts of the five management driver elements of efficiency, creativity, resources, value, and dialectic, and the elements' overall optimization. These five elements, which are detailed below, determine the strategy drivers.

The efficiency-view element is based on the thinking of the transaction cost economics view (Williamson 1975, 1981) and related exchange-efficiency perspectives (Poppo and Zenger 1998; Nickerson and Silverman 2003). In corporate business activities, the perspective of minimizing governance costs (Coase 1993; Demsetz 1988) becomes one of the key elements determining strategy drivers.

The creativity-based view element is inconsistent with the efficiency based on the transaction cost economics view (Williamson 1975, 1981). Companies need to further strengthen the traditional innovative management style of an R&D setup confined within the company to enhance creativity based on the company's own path-dependent knowledge (e.g. Sawhney and Prandelli 2000).

Meanwhile, creativity can be enhanced through collaboration arising from strategic alliances with external partner companies. Companies absorb external partners' ideas and expertise by assessing their own and other companies' knowledge (capabilities) and business models (Chesbrough and Schwartz 2007) through open innovation (Chesbrough 2003) and open business models (Chesbrough 2006), and can also enhance the ingenuity of their own products and services through connect and development strategies (Huston and Sakkab 2006). Absorbing customer competences (Praharad and Ramaswamy 2000; Kodama 2002) also becomes important to strengthening the competitiveness of products and services. Accordingly, creativity becomes an element determining a company's strategy drivers.

The resource-based (Penrose 1959; Chandler 1977; Wernerfelt 1984; Barney 1991) and knowledge-based (Grant 1996; Nonaka and Takeuchi 1995) views are connected to the second element of creativity. A company's organizational boundaries are determined not just by its material assets, but also its intangible knowledge assets, defined as resources, capability, and competences (e.g. Barney 1996; Chandler 1962). Organizational boundaries are not influenced by transaction-cost efficiencies alone. The perspectives of how to acquire knowledge and dynamically allocate and reallocate resources is also important.

---

[1] See Kodama (2009) for details of the five management driver concepts of efficiency, creativity, resource, value, and dialectic. Among these, the "creativity view," especially, promotes the value chain model through the vertical integration of Japanese companies, and reforms existing rules (technology and market) aimed at creating new products and services. The "dialectic view," moreover, synthesizes and integrates diverse knowledge while promoting a co-evolution model aimed at building a new business model among partners crossing industry boundaries.

The value-based view is related to the corporate-value view that queries a company's actions and raison d'etre by asking such questions as "What should we (our company) do?" and "How should we behave?" Organizational identity (Kogut 2000; Thornton 2002; Dutton and Dukerich 1991) forms fixed cognitive frames within an organization, and becomes a key driver for practitioners sharing common meaning and value, and using it as a basis for action (Walsh 1995). Identity, moreover, influences companies' choice of business domains and activities to realize targeted business. Shared organizational identity among practitioners based on corporate vision and mission becomes an important element in building business concepts and determining organizational boundaries (Witt 1998; Penrose 1959).

The dialectic-based view promotes the building of win-win relationships among the stakeholders that comprise the strategy drivers. NTT DoCoMo, for example, built a win-win business model for vertical boundaries comprising mobile phone carriers, manufacturers, content providers, and end users for its mobile Internet services (Kodama 2007a), and produced network externality effects (Bruch and Ghoshal 2004). In Chap. 3, moreover, Sony built a win-win business model (Kodama 2007c) for vertical boundaries comprising game business and console manufacturers, semiconductor manufacturers, game software manufacturers, and end users. The dialectic-based view also becomes the basic framework for a co-evolutionary model that develops targeted business models together with external partners and other stakeholders (e.g. Futuyama and Slatkin 1983).

Companies aim to optimize these management driver elements while determining the strategy drivers of vertical and horizontal boundaries. At this time, companies must review corporate strategy responding to structural changes in the environment and dynamic changes in the competitive environment while adjusting management and strategy drivers to adapt to these changes (see Fig. 7.1).

To further determine a company's optimal strategy and management drivers, practitioners must think and act to expand the dynamic range of knowledge boundaries (see Chap. 6). Thus practitioners must activate the function of "network architecture thinking" (mentioned below in Sect. 7.6), and promote the building of SWS and networked SWS as "network platforms." Then practitioners come and go among the cohesive networks of formal organizations and communities of practice and the sparse networks of SWS and networked SWS, and synthesize the relationships.

Then this practitioners' practice process determines the optimal management and strategy drivers, and promotes boundaries congruence among the elements within the corporate system (strategy, organization, technology, operation, and leadership). The practitioners also simultaneously implement congruence among the environment and the elements that have achieved congruence within the corporate system. The congruence among the elements within this kind of corporate system combines with the dynamic, interactive congruence process among these internal elements and the environment to realize "corporate innovation streams," and results in management innovations for the entire corporate system (see Fig. 7.1).

7.2 Optimizing Organization Architecture for Innovation

While promoting boundaries congruence among the elements within the corporate system (strategy, organization, technology, operation, and leadership) is important to innovational success, building an organizational form within this to achieve congruence of strategy, technology, and operation is an especially important element. Figure 7.2 illustrates the main innovation-oriented forms of organization from the perspectives of operation, leadership, and governance (see Fig. 7.2), and notes key success factors for building optimal organization architecture (OA).

The first organizational form is the R&D model, which is used by major companies in numerous high-tech fields, and mostly exists as independent organizations incorporating research labs and development centers. Many of these specialize in R&D functions, and product and service commercialization is mostly implemented at existing organizations such as operations divisions. Key elements in successful innovation are the presence of the boundary spanner within the company and smooth knowledge transfer to the commercial division. The sustained support of top management is also a factor in leadership and governance.

The second organizational form is the group management model. The cases of NTT-DATA and Sony (see Chap. 3) are leading examples. With this kind of organizational formation, most of the business process operations are implemented by new organizations, and cooperation with the group company stakeholders and collaboration with strategic partners is emphasized. Other key factors in successful innovation include support from parent company on a sustainable resource level, building capital ties with strategic partners, and sustainable collaboration.

The third organizational form is the "integrated organization." The case of NTT-DoCoMo in Chap. 4 is a leading example. This organizational form is an independent model (a formal organization incorporating small-scale organizations including new business development projects) adopted by many major corporations that implements new product, service, and business development within the company. The key to innovative success is to clearly apportion and execute business duties at new and existing organizations, and the most important element is linkage through collaboration among practitioners in each management layer. Collaborating by forming leadership teams at each management level and optimizing business processes through deep collaboration among individual divisions become key factors, especially for major corporations (see Chap. 4).

The fourth organizational form is the "ambidextrous organization." In Japan, the major machine tools corporation Fanuc successfully executed a major technology migration with this form (Shibata and Kodama 2008). Many instances of success through ambidextrous organizations have also been reported at US companies (O'Reilly and Tushman 2004). The key to innovative success is effective resource allocation from top management and collaboration among senior executives.

The fifth organizational model corresponds to provisionally established "mission organizations" such as temporary projects or CFTs. This organizational form is frequently adopted at high-tech companies (including electrical appliances, electronics, IT, and automobiles) in Japan and throughout the world. It resembles the

Organization Architecture	Empirical cases	Features of Organization Architecture			Optimal OA
		Organizational Form	Operation	Leadership & Governance	Key Success Factors
R&D organization	Targets numerous major corporations in high-tech fields	Independent organizations including research labs and development centers	Specializes in R&D functions, and products and services mainly implemented at existing organizations	Support from top management	- Boundary spanner required - Smooth knowledge transfer to implementation division
Group management organization	- Sony, NTT-DATA (Chapter 3) - NTT (Kodama2007b)	Joint venture company with parent company capital governance or with strategic partner	Implements most business process operations at new organization	Cooperation among group companies and collaboration with strategic partners	- Building capital relations and collaborating with strategic partners - Support from parent company
Integrated organization	- NTT DoCoMo (Chapter 4) - NTT (Kodama 2007a)	Integrated organization of independent new and existing organizations within the company	Business allocation and implementation at new and existing organizations	Expand linkage at each management level	- Collaboration through forming leadership teams at each management level - Business process optimization
Ambidextrous organization	- Fanuc (Shibata and Kodama 2008) - US corporations (O'Reilly and Tushman 2004)	Separation of independent new and existing organizations within the company	New organizations become independent from existing organizations and implement business	Expand linkage at top management level	- Effective resource allocation from top management - Collaboration among senior executives
Temporary project, CFT, etc.	-Strategic community project (Chapter 5) - Canon, Sharp (Kodama 2007c) - Honda, Toyota (Kodama 2007c)	Temporarily established mission company	Business allocation and implementation at new organization project and existing organizations	- Expand linkage at each management level - Support from top management	-Collaboration at each management level -Promoting knowledge sharing and transfer with existing organizations - Support from top management

Fig. 7.2 Organization architecture (OA) for innovation

integrated organization in that projects comprising temporary new organizations and business duties at existing organizations are clearly apportioned and implemented. Important factors in innovative success are the promotion of knowledge sharing and transfer among projects and existing organizations through collaboration at each management level, and the continuous support from top management.

Knowledge transfer and sharing across different formal organizations and specialist fields is important to all these organizational forms. Practitioners must flexibly collaborate at each management level through network connectivity action (see Fig. 6.5), and aggressively promote dialog and practice across management levels. This kind of practitioners' "network connectivity action" goes on to promote boundaries congruence among the internal elements (strategy, organization, technology, operation, and leadership) of a corporate system adapted to each organizational form.

Next I will take a closer look at the integrated organizational model, and consider in detail how boundaries congruence is achieved among the corporate system's internal elements.

7.3 Business Architecture for Integrated Organization

In this section, I would like to consider a case of "integrated organization" (see Fig. 7.3) related to business architecture where congruence has been achieved. With companies achieving DoCoMo 's level of integrated organization, it is important to consistently realize optimal business architecture. The ideal strategic organization required of the integrated organization should focus on the integration of heterogeneous organizations with multiple different qualities. This is the aspect of innovative companies merging and integrating, with a regard for balance, the functions of "emergent organizations," which are organizational platforms with emergent (having heterogeneous new knowledge assets) and entrepreneurial elements, with those of "traditional organizations," which have traditional (having a track record and long years of experience) elements (see Fig. 7.3).

Emergent organizations constantly create new knowledge to give rise to new business model (new products, services, and business frameworks) concepts based on imagination and creativity and aimed at innovating in an uncertain environment. Here emergent organizations form SWS and networked SWS incorporating customers and partners, handle knowledge inside and outside the company under a high-risk environment, and go on to promote emergent and entrepreneurial strategies. Individual organizations within the emergent organization take organizational action through the kind of autonomous, distributed leadership that occurs in network and semi-structured organizations (intermediate tight- and loose-coupled organizations) (Brown and Eisenhardt 1998), but constant monitoring from the top ensures overall control of the emergent organization's business direction and objectives. This emergent organization successively gives birth to concepts and prototypes of new products, services, and businesses, and carries out numerous incubations.

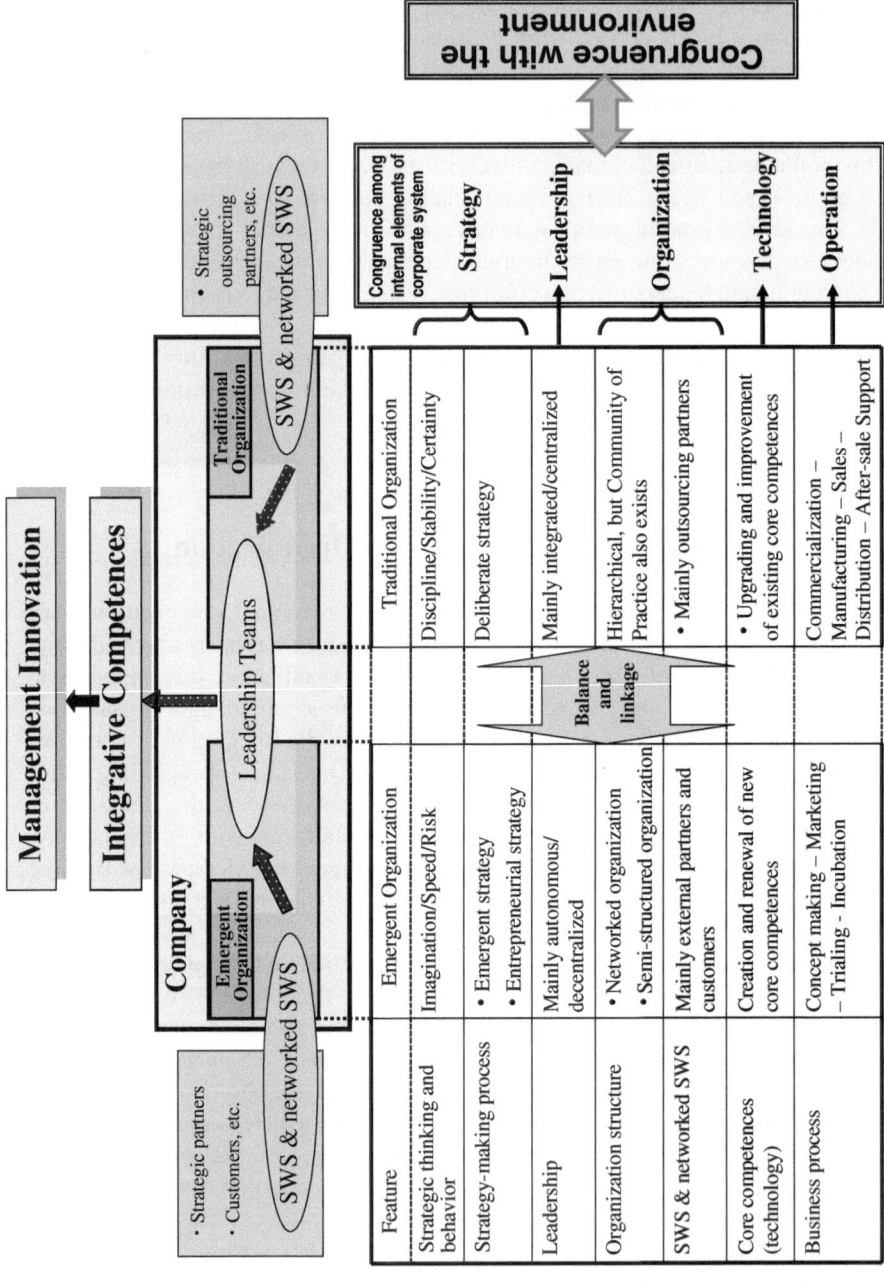

Fig. 7.3 Business architecture for integrated organization

Meanwhile, business processes such as production, sales, distribution, and after-sales support, which are required for the specific launches of these new products and services and to invest, disseminate, and expand in the market efficiently with good timing are important. These business processes are undertaken by the traditional organization platform, and the traditional organization's thoroughly hierarchical command structure promotes routine business based on knowledge accumulated over many years. Traditional organizations adopt strategy proposals after rules-based deliberations from closely regulated leadership, raise business process efficiency and carry out incremental improvements, and promote best practice at the workplace through communities of practice. Outside the company, strong-tie SWS and networked SWS form with strategic partners having either short- or long-term dealings with the company. Traditional organizations then become the focus for quick, efficient, and reliable market investment, distribution, and expansion of the results of the emergent organizations' innovative new product and service concepts.

These two general classifications of organization possess "paradoxical" elements whereby emergent organizations pursue creativity and independence and traditional organizations efficiency and regulation. The two organizational models constantly engage in tugs-of-war and conflict within the company, which obstruct even as knowledge is merged and integrated. Promoting this integration is the task of the "leadership teams" (case studies appear in this book).

"Leadership teams" are formed by those in charge (president, senior executive managers, executives, divisional heads, project leaders, and others) at each management level of the emergent and traditional organizations, including top, middle, mixed top and middle, cross-functional, and task force. The teams merge and integrate knowledge from emergent and traditional organizations, and take on the role of creating integrative competence for the entire company (Kodama 2007a). To achieve integrative competence, leadership teams consciously implement integrative strategy based on disciplined imagination, thus integrating paradoxical strategies—creative and emergent with deliberate and analytical. As well as integrating strategies, leadership teams[2] at emergent and traditional organizations have a mission to implement balance and linkage among the individual management elements of organization, technology, operation, and leadership.

Leadership teams promote thorough understanding of problems and issues through dialectic and creative dialog among leaders, and each leader acknowledges the role and value of others' work through mutual communication and collaboration. Accordingly it is important that leaders transform the various conflicts arising

---

[2]This is also an ideal system of business architecture for integrated organizations. With current corporate activity, however, optimizing business architecture entails various problems. Prominent among them is business-related intrusion and cannibalization among multiple emergent organizations at major corporations, and the impact of the power imbalance on business architecture congruence among emergent and traditional organizations. This kind of non-conformance of business architecture surfaces when a company faces new environments. Thus the degree of business architecture congruence can also influence corporate performance. There is no space to expand on the topic in this text, but I would like to present the results of my research in the future.

among themselves into "constructive conflict." Within the leadership teams, every leader is required to think and act in response to such questions as "What actions should I undertake, with what strategies and tactics, to contribute to corporate growth and innovation?" when considering their mission targets. Meanwhile it is important that the president, as final decision-maker at the top of the leadership teams, exhibits top-down leadership as required while strengthening the collaborative links between president and leaders by aggressively generating dialog and debate within leadership teams and maximizing leadership coherence at each management level. Then the leadership teams build optimal business architecture as a corporate system aimed at the business achievements of short- and long-term innovation. Thus the integrated synergy effects of leadership from individual leaders enables the creation of strong new business value chains and new customer values.

Leadership teams correspond to layered SWS and networked SWS within the company. Practitioners at integrated organizations not only practice business at formal (emergent and traditional) organizations, but also solve problems and search for new business ideas[3] aimed at achieving strategic objectives through practice at the SWS and networked SWS leadership teams formed from each management layer. Looking at the temporal axis of strategy practice, practitioners implement "network connectivity action" through traffic and synthesis among the kind of dual networks shown in Fig. 6.5. Thus integrated organizations are also the "process-based organization" of Fig. 6.4.

The process-based organization aims to achieve congruence among the individual elements of strategy, organization, technology, operation, and leadership for such emergent and traditional organizations as well as mutual congruence of the management elements at these organizations, thus maintaining mutual linkage and balance. At the same time, integrative organizations dynamically transform these management elements in response to environmental change, and promote boundaries congruence among management elements and among the two organizations (see Fig. 7.3). This kind of dynamic boundaries congruence process goes on to achieve optimal business architecture for the integrated organization.

7.4 New Knowledge Creation From the Process-Based Organization

SWS and networked SWS at the process-based organization enable interaction of dialog and practice among practitioners, and exist as informal organization models for creating new knowledge. SWS and networked SWS are not opposed to formal organizations, however. Rather, the process-based organization embraces management techniques acquired from the management of routine and daily tasks at formal

---

[3] In network theory, "leadership teams" are small-world structures (SWS) arising from shortcuts among actors within and among organizations. See, for example, Paduda (2008).

organizations focused on problem-solving or creatively oriented tasks at SWS and networked SWS.

Panasonic's "flat and web organization" (Kodama 2007a) can be described as a twenty-first century organizational model similar to that of a process-based organization. The "web" of "flat and web" means that transmission and sharing of information within and among organizations is implemented in an open environment such as the World Wide Web (WWW), and suggests the existence of various multilayered SWS and networked SWS within and among individual organizations. These SWS and networked SWS create new meanings and dynamic contexts in interactive information and knowledge within and outside organizations, including among stakeholders, and become a source of knowledge creation.

SWS and networked SWS ensure creativity and flexibility when expressing corporate vision and strategic objectives in terms of specific work duties. SWS and networked SWS may also exist in multilayered, stratified structures (Kodama 2007b, 2007c), and are one means of sharing and creating contexts with the key people attached to the various layers inside and outside the organization in order to achieve efficient and speedy creativity. The formation of SWS through short cuts and rewiring is also one means of forming new contexts and knowledge. Meanwhile flat, layered organizations ensure speed and efficiency when specifically implementing work duties.

In order to adapt to environmental change (or create a new environment), the key is either to divide business into flat formal organizations and SWS/networked SWS as the situation requires, or to flexibly and spontaneously change the organizational structure through simultaneous implementation and integration. The process-based organization simultaneously pursues existing business with the objective of creating short-term profit (environment adaptive strategy) and the innovation of long-term business creation (environment creation strategy) by systematically enhancing efficiency, and goes on to realize "corporate innovation streams." The practitioners' network architecture thinking (mentioned later) is an important perspective when building a process-based organization. This thinking goes on to optimize business architecture and realize corporate innovation streams.

7.5 Network Architecture Thinking

I would like to use the case studies in this book to point out the significance of strategy practice resulting from network architecture thinking as an organizational behavior that is common among successfully innovative companies. This network architecture thinking becomes an enabler that integrates the heterogeneous knowledge required for strategy and technology transfers, and builds business architecture congruent with environmental change.

The formal organizations and SWS/networked SWS at process-based organizations are also knowledge platforms for practitioners that share dynamic context (time, place, and human relationships) and create new knowledge. The process-based organization moreover, is a time-space continuum where practitioners interact to share context with others, and transform this context to create and change new knowledge. This sharing of tacit knowledge and the time-space continuum of dialog and practice make up the process-based organization. In this section, I will consider how to build a process-based organization from a practical viewpoint. I believe that describing the organizational mechanisms centered on the process-based organization is important both from an academic and a practical viewpoint.

The organization and the individual are dialectically related. Practitioners change organizations through the human power possessed by individuals and the practical consciousness of tacit knowledge relating organizationally and cyclically to the organization within the time-space continuum and dynamic context of the "here and now" as a process-based organization. People possess the power to try to change the content of the organization through their own actions, even as they accept the organizational restrictions that they themselves created (Giddens 1984). In the process-based organization, the network platforms that create bridges between these individuals and organizations (companies) are the SWS and networked SWS, and the micro-presence of individuals influences the macro-structure of organizations, companies, industries, and society as a whole by forming SWS and networked SWS. Accordingly, important points for analysis include the key position of SWS and networked SWS as a micro–macro linkage of social networks; how individuals manage the SWS and networked SWS, form and accumulate social capital, and influence corporate performance amid the relationships among individuals, organizations, SWS and networked SWS, companies, and industries; and conversely, what influence SWS and networked SWS have on individuals. Considered from this viewpoint, SWS and networked SWS is an important unit of analysis.

Seen from the viewpoint of knowledge management, meanwhile, creation of new knowledge capital centers on SWS and networked SWS, which are also important from the perspective of clarifying the process by which diverse knowledge crosses SWS/networked SWS boundaries and becomes integrated. The point of forming SWS/networked SWS to create new knowledge also has practical significance for the practitioners.

A common point arising from this book's case analyses is that multilayered SWS and networked SWS always exist inside and outside the company. This is because practitioners have created and consolidated by subjectively influencing the environment (including customers) and others inside and outside the company. Sony, NTT-DATA, DoCoMo, Honda, Toyota, Takara, Omron, and other companies have formed numerous SWS and networked SWS inside and outside their own companies and developed these multilayered networks. When practitioners form SWS and networked SWS from network architecture thinking, how the topology will be? Here, I would like to mention the two focal points of internal and external SWS architecture as patterns of network architecture observed in my previous corporate research.

7.6 Network Architecture of the Small-World Structure

First, I want to reconfirm the definition of an SWS (a small-world structure). The small-world networks mentioned by Watts and Strogatz (1998) in the field of network theory meant networks capable of reaching from certain nodes to certain other nodes via a small number of connections, and also certain nodes connecting with each other. Technically, such structures are defined as "networks with a short average route but a high cluster coefficient," and in typical small-world networks, links between distant nodes and random shortcuts (or rewiring) exist within networks systematically linking adjacent nodes (with individuals in the community as connecting points) (see Fig. 7.4).

Mentioning points of difference between the formation of a regular network and a random network, Watts and Strogatz (1998) demonstrated that small-world networks fall into a category between the two with respect to their degree of randomness. On the one hand, each node in a small-world network is embedded in a local cluster, and thus the small-world network possesses a higher clustering coefficient than does the random network. On the other hand, short paths that directly connect two nodes located spatially distant from each other (a characteristic typically seen in random networks) are also present in small-world networks. These shortcuts (or rewiring) provide easy access to most of the nodes in the network,

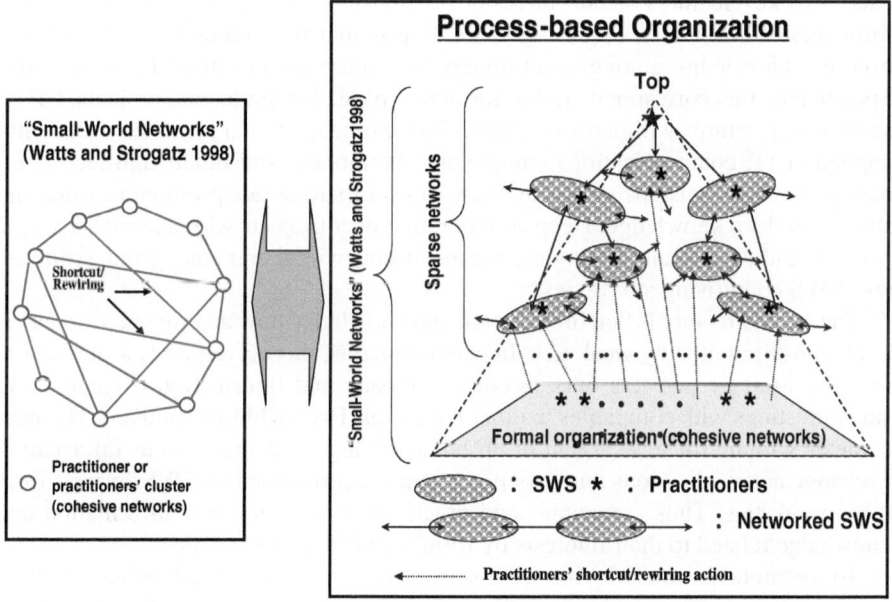

Fig. 7.4 Small-world networks (Watts and Strogatz 1998) and the process-based organization

while local clustering tends to foster densely reliable accessibility. Small-world networks are more centralized than random networks due to the high number of local clusters, but not at the level inherent to centralized networks.

Considered from such a viewpoint, the process-based organizations comprising sparse networks (equivalent to SWS and networked SWS) created from numerous practitioner short-cuts and formal organizations of multiple-cluster cohesive networks are equivalent to the "small-world networks" mentioned by Watts and Strogatz (1998) (see Fig. 6.4). In order to analyze in detail the practitioners' practice process in this book, however, these networks are differentiated as sparse and cohesive from thinking that emphasizes different business contexts inside and outside the organization, and interdependently among practitioners, and sparse networks are termed SWS (small-world structures) (see Fig. 7.4). The reason is that not only the business context but also the strategy practices of practitioners at the sparse networks of SWS/networked SWS and cohesive networks of formal organizations and communities of practice in individual networks differ (illustrated in Fig. 6.4). Next, I would like to consider SWS network architecture of sparse networks formed from these practitioner short-cuts.

7.6.1 Internal Network Architecture of SWS

New clusters form at the sparse networks comprising multiple practitioner short-cuts. Two key features stand out about relationships among the practitioners that form these SWS clusters (see Fig. 7.5). One is that the mutual flow of information and knowledge among practitioners is a relationship united by strong ties. Specifically, this corresponds to the formation of leadership teams, projects, CFTs, and other structures within the company. These organizational forms are frequently applied at IT, communications equipment, automobile, and other high-tech companies. The relationships united by these kinds of strong ties promote transfer and sharing of deep knowledge. Deep collaboration over projects with external strategic partners and large-scale development consortiums within and among industries are also SWS comprising strong ties.

The second feature is that information and knowledge flow among practitioners is a relationship united by weak ties. In concrete terms, this corresponds to the search for new strategic partners outside one's company and information exchange and study meetings with companies in other industries. Even within a company, business contexts with no (or few) urgent or problem-solving elements, such as information exchange across divisions and group companies, correspond to SWS formed from these weak ties. Thus companies and practitioners scan for new information and knowledge related to their interests by forming SWS with weak ties.

To promote sustainable innovation activities, the dual relationship of these strong and weak ties must be mutually complementary and continually synthesized. Specifically, companies (practitioners) adapted to such areas as environmental, strategic, and technological change must dynamically change the qualities of the sparse networks formed from flexible short-cuts (see Fig. 7.5). When external

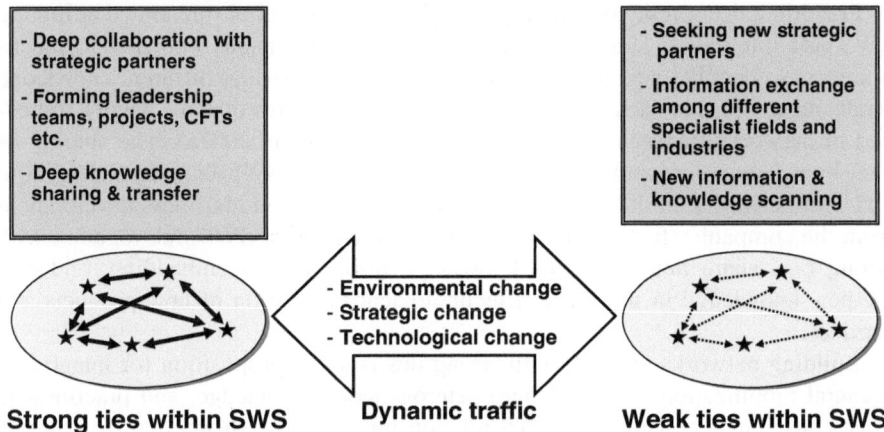

Fig. 7.5 SWS Internal network architecture: strong and weak ties

information and knowledge scanning turns up prominent potential partners, for example, companies must transform weak-tied SWS to strong-tied SWS and build firm strategic partnerships by exchanging new contracts. Conversely, to adapt to new business opportunities and technology changes, it is vital that a company reconsiders strong-tie relationships with existing strategic partners and converts them to weak-tied relationships.

7.6.2   The External Network Architecture of SWS

7.6.2.1   SWS Relationships

Recently new product and service development in high-tech fields has been pressured by the need to merge and integrate different technologies. Technological innovation in the past sought out, pursued, and developed specialist knowledge, but now numerous cases have arisen where technologies in certain fields involved in new product development based on unprecedented new ideas come to be realized by merging with technologies in other fields. How to somehow integrate diverse, distributed knowledge is a key questions (from a technology viewpoint, this is the question of how to integrate knowledge from different technology fields). Distributed new knowledge is embedded in SWS and networked SWS distributed in the space-time continuum. To integrate knowledge, individual knowledge items must transcend SWS boundaries and accumulate in networks. Thus it is necessary to connect distributed SWS on networks, and to deeply embed in networks the knowledge distributed in each SWS. In social network theory, SWS can also be described as cliques and clusters (gatherings of practitioners linked with strong or weak ties) of practitioners, and the connection of SWS to other SWS via networks (networked SWS) correspond to "ties."

Practitioners committed to multiple SWS take on a central role aimed at linking SWS and integrating knowledge. To integrate heterogeneous knowledge, practitioners must deeply understand and share (knowledge sharing) different knowledge (tacit and explicit) in each SWS, and this shared knowledge must be deeply embedded in networks transcending SWS boundaries (Kodama 2007a). The sharing of tacit knowledge, especially, requires strong ties linking SWS that share deep context over networks. In the cases of new product, service, and business development from the companies featured in this book's case analyses, SWS link together with strong ties, share heterogeneous knowledge through deep embedding, and create new knowledge in the form of technological integration of new products and services.

Building networks of SWS with strong ties is a key proposition for integrating (general mobilization of knowledge) heterogeneous knowledge, and practitioners must consciously consider the SWS relationships of these strong ties. To embody new concepts such as new business models, practitioners must promote technology integration through strong-tied SWS networks (networked SWS). The SWS might also be reorganized to adapt to dynamic environmental change and pursue the desired technology development. It is also important to embed expertise within strong-tied SWS networks as tacit knowledge acquired from practitioners through reflective practice.

Meanwhile, social network theory tells us that weak ties are likely to provide a bridge to new heterogeneous information (Granovetter 1973). It also suggests that the weak ties with the "structural hole" mentioned by Burt (Burt 1992) are highly likely to enable practitioners to access new information and grasp new business opportunities. This book is especially concerned with the search for an effective method for building SWS networks with weak ties (see Fig. 7.6). While corporate research in this area is scant, the case of NTT DoCoMo described in Chap. 4 is applicable here. The i-mode business model arose through integration of diverse knowledge, including mobile phone, technology platform, and content development. Within this model, mobile phone and technology platform development is generated by various SWS inside and outside the DoCoMo organization, and linked with "strong ties."

Meanwhile, with regard to the development of diverse content (encompassing a wide range of text content, games, positioning data, music and image distribution, two-dimensional bar codes, and e-money services such as FeliCa), DoCoMo not only built close network connections with specific content providers, but also built SWS networks with content providers to which it had weak ties, and targeted the opportunity to access new content information and use it to create new content business.

As I have shown above, when designing SWS networks, an important proposition for promoting new business is to build SWS networks with weak ties alongside those with strong ties, and practitioners must deliberately consider and synthesize the relationships between these strong- and weak-tie SWS (see Fig. 7.6). This maintains strong ties while forming further weak-tie SWS networks, bridging the structural

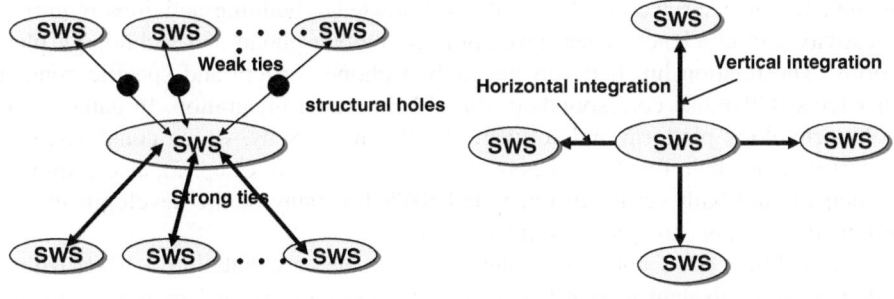

Relationships between SWS   **Structural nature of SWS**

Fig. 7.6 SWS external network architecture: relationships and structural nature

hole at the right time, and absorbing and integrating heterogeneous knowledge. The i-mode business model was realized through the skilful use of "SWS architecture."

7.6.2.2 The Structural Nature of SWS

In order to rapidly create, disseminate, promote, and expand high-quality knowledge, practitioners need to consider elements both of SWS relationships and structural nature. The structural nature of SWS can be divided into vertical and horizontal integration of the SWS.

SWS vertical integration creates SWS hierarchies. These occur when practitioners at various sections within the company form a number of cross-functional SWS while forming layered structures of these SWS at management level. These hierarchies are observed, for example, in cases of large-scale new products development, where professionals in various divisions and specialist fields related to the product targeted for development, such as design of the overall architecture and each subsystem, software and hardware development, and production cooperate, form "ba" at each management level, and create hierarchical structures. This kind of vertical integration of SWS is especially effective in new product development processes that require merging and integration of various technologies (Kodama 2007b). Autonomous SWS formed from professional groups with regard to SWS vertical integration ensure creativity and flexibility when presenting specific work duties for the mission of new product development. Meanwhile, the SWS hierarchy guarantees efficiency and swift decision-making when executing those duties.

This SWS vertical integration is also closely connected to the industry value chain that determines the vertical boundaries of corporate and industrial strategy drivers. For example, Japanese mobile phone carrier NTT DoCoMo (see Chap. 4) displayed leadership (with both power and initiative) with regard to numerous vendors (including mobile handset development and production manufacturers and semiconductor chip manufacturers), built up SWS vertical integration, and maintained new service development and knowledge-building activities requiring creativity and efficiency when developing its mobile phone and technology platforms. The relationship between the mobile phone carriers and specific content providers (CPs) also corresponds to this SWS vertical integration. In games console technology platform and software development, Sony, which constructed an industry value chain for the games business (mentioned in Chap. 3), also exhibited leadership and built vertically integrated SWS for many of the development and production vendors and games software houses.

Meanwhile, "SWS horizontal integration" creates a collaborative SWS with a relationship equivalent to that between customers and external partners, and has few of the hierarchical elements seen in SWS vertical integration. This is a case of sharing and disseminating common knowledge with partners, and carrying out new product and service development with an equal relationship. NTT-DATA's joint venture strategy formed through strategic alliances with partners (see Chap. 3), and the university, hospitals, private businesses and non-profit organisations have worked together to advance virtual networking in the field of veterinary medicine (see Chap. 5), for example, which correspond to this SWS horizontal integration. Furthermore, DoCoMo exploits SWS horizontal integration to widely disseminate knowledge of the i-mode and FOMA 3G mobile phone system created and accumulated within DoCoMo to overseas communications carriers. DoCoMo has also targeted this SWS horizontal integration by collaborating with other fixed communications carriers (including the NTT group) and Internet Service Providers (ISPs). DoCoMo was able to exhibit integrative competences by synthesizing the dual aspects of this SWS horizontal integration and the above-mentioned SWS vertical integration to ascertain a yield increase from the synergistic effects of future broadband and mobile phone businesses (see Fig. 7.6).

From the above discussion of SWS internal and external architecture, and considered from the practitioners' viewpoint, the question arises how practitioners should realize SWS architecture creating new knowledge. Practitioners inquire about the specific topics they face each day, asking "How can we enable access to new knowledge? "How can we create new business ideas?" and "At what level should we keep the cooperative relationships with our partners?" It is important that the practitioners, without needing to grasp the existing conceptual paradigm, deliberately and subjectively create and consolidate potential SWS networks, including partners and customers, that are the source of the knowledge contexts hidden in the background of existing formal organizations, companies, and industry structures and among different industry types, and focus their own thoughts and actions on the internal and external architecture of SWS.

Companies (practitioners) in business domains with highly uncertain, dramatically changing environments need to form (and rebuild as required) dynamic SWS networks considering SWS internal and external architecture, and create new targeted knowledge. Dynamic processes that constantly and deliberately promote SWS networks within corporate organizations and among partners (including customers) become the drivers creating the corporate innovation streams of incremental and radical (discontinuous) innovation.

References

Barney, J. (1991). Firm resources and sustained competitive advantage. *Journal of Management*, 17(3), 99–120.

Barney, J. (1996). *Gaining and Sustained Competitive Advantage*. Boston, MA: Addison-Wesley.

Brown, S., Eisenhardt, K. (1998). *Competing on the Edge*. Boston, MA: Harvard Business School Press.

Bruch, H., Ghoshal, S. (2004). *A Bias for Action: How Effective Managers Harness Their Willpower, Achieve Results, and Stop Wasting Time*. Boston, MA: Harvard Business School Press.

Burt, S. (1992). *Structural Holes: The Social Structure of Competition*. Cambridge, MA: Harvard University Press.

Chandler, A. D. (1962). *Strategy and Structure: Chapters in the History of American Enterprise*. Boston, MA: MIT Press.

Chandler, A. D. (1977). *The Visible Hand: The Managerial Revolution in American Business*. Cambridge, MA: Belknap Press.

Chesbrough, H. (2003). *Open Innovation*. Boston, MA: Harvard Business School Press.

Chesbrough, H. (2006). *Open Business Models: How to Thrive in the New Innovation Landscape*. Boston, MA: Harvard Business School Press.

Chesbrough, H., Schwartz, K. (2007). Innovating business models with co-development partnerships. *Research Technology Management*, 50(1), 55–59.

Coase, R. H. (1993). The nature of the firm: influence. in *The Nature of the Firm*. O. E. Williamson and S. G. Winter (eds),. New York: Oxford University Press, pp. 61–74.

Collins, J., Porras, J. (1994). *Built to Last: Successful Habits of Visionary Companies*. New York: Harpercollins.

Demsetz, H. (1988). The theory of the firm revisited. *Journal of Law and Economic Organization*, 4, 141–161.

Dutton, J. E., Dukerich, J. M. (1991). Keeping an eye on the mirror: image and identity in organizational adaptation. *Academy of Management Journal*, 34(3), 517–554.

Futuyama, D., Slatkin, M. (1983). *Coevolution*. Sunderland, MA: Sinauer Associates.

Giddens, A. (1984). *The Constitution of Society*. Berkeley. CA: University of California Press.

Granovetter, M. (1973). The strength of weak ties. *American Journal of Sociology*, 78(6), 1360–1380.

Grant, R. (1996). Toward a knowledge-based theory of the firm. *Strategic Management Journal*, 17(Winter Special Issue), 109–122.

Hamel, G., Prahalad, C. K. (1989). Strategic intent. *Harvard Business Review*, 67(3), 139–148.

Huston, L., Sakkab, N. (2006). Connect and develop inside Procter & Gamble's new model for innovation. *Harvard Business Review*, 84(3), 58–66.

Kodama, M. (2002). Strategic partnership with innovative customers: a Japanese case study. *Information Systems Management*, 19(2), 31–52.

Kodama, M. (2007a). *The Strategic Community-Based Firm*. London: Palgrave Macmillan.

Kodama, M. (2007b). *Knowledge Innovation – Strategic Management as Practice*. Cheltenham: Edward Elgar Publishing.

Kodama, M. (2007c). *Project-Based Organization in the Knowledge-Based Society*. London: Imperial College Press.

Kodama, M. (2008). *New Knowledge Creation Through ICT Dynamic Capability-Creating Knowledge Communities Using Broadband*. Charlotte, NC: Information Age Publishing.

Kodama, M. (2009). *Innovation Networks in Knowledge-Based Firm – ICT-Based Integrative Competences*. Cheltenham: Edward Elgar Publishing.

Kogut, B. (2000). The network as knowledge: generative rules and the emergence of structure. *Strategic Management Journal*. 21, 405–425.

Nickerson, J. A., Silverman, B. S. (2003). Why firms want to organize efficiently and what keeps them from doing so: Inappropriate governance, performance, and adaptation in a deregulated industry. *Administrative Science Quarterly*, 48(3), 433–465.

Nonaka, I., Takeuchi, H. (1995). *The Knowledge-Creating Company*. New York: Oxford University Press.

O'Reilly, C., III, Pfeffer, J. (2000). *Hidden Value: How Great Companies Achieve Extraordinary Results with Ordinary People*. Boston, MA: Harvard Business School.

O'Reilly, C., III, Tushman, M. (2004). The ambidextrous organization. *Harvard Business Review*, 82, 74–82, April.

Paduda, G. (2008). Enhancing the innovation performance of firms by balancing cohesiveness and bridging ties. *Long Range Planning*, 41(4), 395–419.

Penrose, T. (1959). *The Theory of the Growth of the Firm*. New York: Wiley.

Peters, T., Waterman, R. (1982). *In Search of Excellence*. New York: Harper & Row.

Poppo, L., Zenger, T. (1998). Testing alternative theories of the firm: transaction cost, knowledge-based, and measurement explanations for make-or-buy decisions in information services. *Strategic Management Journal*, 19(9), 853–877.

Porter, M. (1985). *Competitive Advantage*. New York: Free Press.

Prahalad, C. K., Ramaswamy, V. (2000). Co-opting customer competence. *Harvard Business Review*, 78(1), 79–87.

Santos, M., Eisenhardt, K. (2005). Organizational boundaries and theories of organization. *Organization Science*, 16(5), 491–508.

Sawhney, M., Prandelli, E. (2000). Communities of creation: managing distributed innovation in turbulent markets. *California Management Review*, 42, 24–54.

Shibata, T., Kodama, M. (2008). Managing technological transition from old to new technology: case of Fanuc's successful transition. *Business Strategy Series*, 9(4), 157–162.

Thornton, P. (2002). The rise of the corporation in a craft industry: conflict and conformity in institutional logics. *Academy of Management Journal*, 45(1), 81–101.

Walsh, J. P. (1995). Managerial and organizational cognition: notes from a trip down memory lane. *Organization Science*, 6(3), 280–321.

Watts, J., Strogatz, S. (1998). Collective dynamics of "small-world" networks. *Nature*, 393(4), 440–442.

Weick, K. E. (1995). Sensemaking. *Organizations*. London: Sage.

Wernerfelt, B. (1984). A resource-based view of the firm. *Strategic Management Journal*, 5(1), 171–180.

Williamson, E. (1975). *Markets and Hierarchies: Analysis and Antitrust Implications*. New York: Free Press.

Williamson, O. E. (1981). The economics of organizations: the transaction cost approach. *American Journal of Sociology*, 87(3), 548–557.

Witt, L. A. (1998). Enhancing organizational goal congruence: a solution to organizational politics. *Journal of Applied Psychology*, 83, 666–674.

Chapter 8
Conclusion

8.1 The Dynamics of Business Architecture

The essence of strategic management goes beyond companies creating appropriate future-oriented strategies while adapting to environmental change. It also involves optimizing each of the corporate system management components (organization, technology, operation, and leadership) in line with these strategies and dynamically and integratively developing them to realize corporate continuance and growth. An important issue from the implementation aspect is how companies can dynamically change corporate boundaries and adapt to the environment (or create new environments) while considering congruence with the environment. To achieve these aims, companies must determine the strategy objectives of sustainable, competitive products, services, and business models, and achieve them by implementing optimal design of vertical (value chains to realize corporate strategy objectives) and horizontal (expansion and diversification of the business domain) boundaries. In this book I have used the term "business architecture" for the optimal design of a corporate system comprising the management elements of strategy, organization, technology, operation, and leadership in order to design corporate strategy compatible with the environment, and suggested a basic theoretical framework through multiple in-depth case studies.

The concept of business architecture also involves optimizing corporate boundaries aimed at realizing targeted business models and optimal corporate system design ideas incorporating stakeholders. Optimizing the corporate system requires partial and overall optimization of the individual management elements of strategy, organization, technology, operation, and leadership—the elements comprising the corporate system that has achieved congruence with the environment. The management that optimizes these corporate boundaries and corporate system elements is "boundary management," which I have described in this book.

The concept of "boundaries congruence" inside and outside the corporate system is key to realizing dynamic strategic management. Building optimal architecture for the environmental change and the management elements of strategy, organization, technology, operation, and leadership is also important. In this book, based on numerous corporate case studies, I have analyzed the concept and optimization processes of business architecture as a corporate system.

Existing research describes "business architecture" as a structure that integrates business concepts and combines various management activities as targets of corporate strategic behavior, including relationships with stakeholders, products and services, organizations, business processes, and ICT. Thus the structure involves interdependence and relationships among elements of management activity. The features of "business architecture" in this book, however, extend the concept beyond the conventional definition of product and process architecture (a concept regarding architecture comprising products and services, and the business process mechanisms required to realize it) for IS research and product innovation. The expanded concept considers congruence of corporate environment change and congruence among management elements inside the corporate system as corporate architecture while embracing dynamic congruence on a time-space continuum.

The dynamics of product and process architecture adapted to technological innovation also influence business architecture. Conversely, the dynamics of business architecture create changes in product and process architecture. In a dynamically changing environment, companies must dynamically structure and restructure not only product system architecture and business processes, but also business architecture as optimal architecture for the corporate system. In this book, I have suggested that the optimal design of business architecture as an optimal corporate system has promoted congruence with the environment, and been a key element in sustainably implementing corporate innovation streams.

8.2 Innovation from Network Architecture Thinking

This book has also noted the importance of practitioners' network architecture thinking through numerous case analyses relating to business architecture as a corporate system for realizing new products, services, and businesses. Based on dynamically changing environments, companies must dynamically build and rebuild product systems; business processes; and optimal business architecture as a corporate system through network architecture thinking.

In the twenty-first century, diverse innovative technologies are transforming conventional business structures in every industry (see Fig. 8.1). The business models of the twentieth century concentrated a company's resources in its own core competences, mostly within the paradigm of mass production and sales, and delivered standard products and services to pursue economies of scope and scale. Companies also consistently formulated and implemented strategy for the predictable conditions of a slowly changing environment based on closed, bureaucratic, layered organizations. Now, however, technological innovation led by ICT is increasingly transforming this environment, and will continue to do so in the future.

In the future, every kind of industry and company must develop personalized business adapted to mass-customization, diverse business solutions, and individual

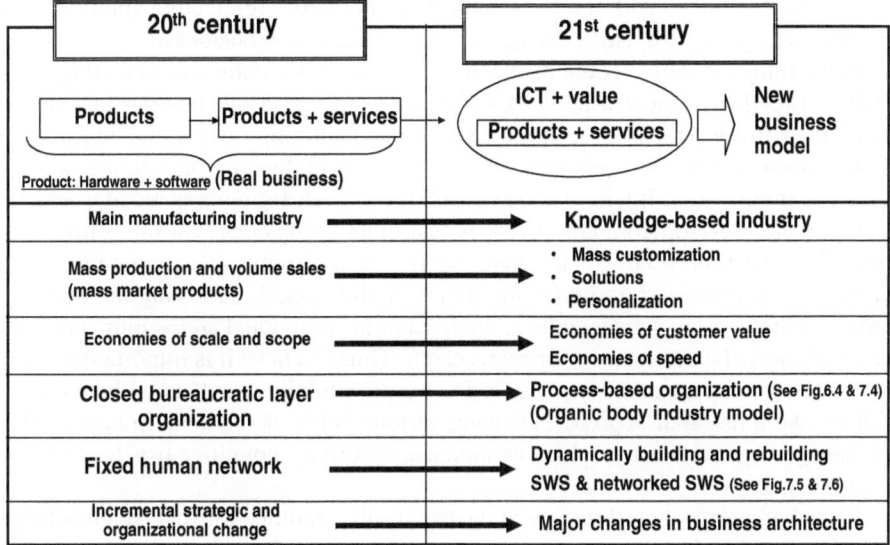

Fig. 8.1 The development of business formations

customer needs as new business models exploiting ICT and other high-tech technologies, and create new knowledge and customer value. Thus every industry will be able to assert that it is a "knowledge-based industry." These knowledge-based industries require economies of customer value and speed, and achieving new innovations becomes an urgent business issue. The corporate model template for such knowledge-based industries is the "process-based organization" arising from network architecture thinking considered in this book. It is a corporate system as an "organism" that is able to flexibly build and rebuild the SWS and networked SWS of sparse networks and the formal organizations and communities of practice of cohesive networks adapted to major environmental change.

These days the outstanding core technologies and business model ideas in pioneering technologies and service fields such as electronics, consumer electronics and communications equipment, machine tools, ICT, e-business, content, semiconductors, biotechnology, and automobiles have spread around the world and innovate on a daily basis. The combination and integration of knowledge distributed worldwide will surely create the new, dynamic business architecture that will create the business processes and models to realize future new products and services. By continuously setting out new product, service, business and other innovations, many of the companies are already finding it difficult to retain sole control over innovations made under the conditions of the hierarchical organization and autonomous corporate system of the mass-production era. Because of this, organizational and human network formations involving various stakeholders are important. The building of "business architecture" as an optimal corporate system resulting from the process-based organization incorporating the SWS and networked SWS formations, posited in this book, will surely become key "network architecture thinking" for practitioners growing existing business while creating new business.

In the future, it will become increasingly important for companies creating competitive excellence under a network economy to integrate outstanding knowledge distributed inside and outside the organization (including customers) in an open system from multiple perspectives, and to manage with the aim of building optimal business architecture. It will also be essential, however, for companies to gradually maintain and develop core competences that other companies will find difficult to copy. The question of how distributed, heterogeneous knowledge can be integrated in this way is a research topic for the future. In this regard, clarifying the dynamic process at a microlevel (especially at the level of the individual and among individuals) will surely be the subject of new research. Thus I believe it is important to form new foci and interpretations with regard to people's thinking and action boundaries, and to take a research approach crossing various fields of scholarship beyond that of management studies to include economics, sociology, cognitive psychology, and network theory.

In the knowledge-based society of the twenty-first century, the diverse knowledge held by individuals becomes the creative source for valuable products, services, and business models conferring a new competitive power. The business architecture mentioned in this book can be considered as a new insight with value for numerous practitioners aiming to achieve innovation.

Appendix: Research Methodology and Data Collection

I adopted a qualitative research methodology due to the need for rich data that could facilitate the generation of theoretical categories I could not derive satisfactorily from existing theory. In particular, due to the exploratory nature of this research and my interest in identifying the main people, events, activities and influences that affect the progress of innovation, I selected the grounded theory-based study of data interpretation, which was blended with the case study design and with ethnographic approaches (Locke 2001).

The research data came primarily from longitudinal study during a 16-year period (1993–2008) examining new knowledge creation process with respect to new products and services development at a large company in competitive ICT and other high-tech fields. This research paradigm, which was based on in-depth qualitative study, has some similarity to ethnography (Atkinson and Hammersley 1994) and other forms of research (Lalle 2003) that derive their theoretical insights from naturally occurring data including interviews or questionnaires (Marshall and Rossman 1989). Especially, the author of this paper himself serves as a project manager of new product development in NTT and NTT DoCoMo, Japan's largest telecommunications companies. This experience provided the author with direct knowledge and detailed information with which the accuracy of the empirical analyses in this research was enhanced. Research data and insight are gained alongside or on the back of the intervention.

The data collected over the ten years of the intervention have derived from work with practitioners involved in a large number and variety of customers and outside partner as well as internal organization members. During these interventions, the expressed experiences, views, actioncentered dilemmas, and actual actions of participants were recorded as research data in a variety of ways, including notes, internal and outside rich documents, etc. The theory that has emerged from this research has centered around the concepts of new knowledge creation process collaborated with outside partners and customers.

The data analysis for the research consisted of three stages: (1) developing in-depth case history of a big project's activities from the raw data that I could gain all the information, (2) open coding and subsequent selective coding the in-depth case history for the characteristics and origin of knowledge creation process, and (3) analyzing the pattern of relationships among the conceptual categories.

M. Kodama, *Boundary Management*, DOI 10.1007/978-3-642-03789-4,

In the first stage of the data analysis, I constructed chronological descriptions of the project's activities with respect to knowledge creation process, describing how it came about, when it happened, who was involved, and major outcomes. Through this work, I completed an in-depth case history of the project.

The second stage of analysis involved coding the in-depth case history with respect to its characteristics, origin and effects. This was a highly iterative procedure that involved moving between the in-depth case history, existing theory, and the raw data (Glaser and Strauss 1967). Data were subjected to continuous, cyclical, evolving interpretation and reinterpretation that allow patterns to emerge.

The grounded theory approach was adopted based upon the researchers' interpretation and description of phenomena based on the actors' subjective descriptions and interpretations of their experiences in a setting (Locke 2001). This "interpretation of an interpretation" strives to provide contextual relevance (Silvermann 2000). From the in-depth case history, I initially advanced first-order descriptions based on broad categories that were developed from the existing theory, and then refined these categories by tracing patterns and consistencies (Strauss 1987). The analysis continued with this interplay between the data and the emerging patterns until the patterns were refined into conceptual categories (Eisenhardt 1989). The third stage of data analysis was to examine the empirical in-depth case results across the selected categories and the theoretical literature, and to develop the logic of the conceptual framework and generate new theory.

NTT-Data, Sony, Toyota, Honda, Omuron, Takara, Panasonic and Cannon case studies are based on in-depth interviews with many senior managers of corporations, and on internal and external materials. From 2000 to 2003, the author worked as a project leader in product development at NTT DoCoMo. During this period, he gathered data through informal dialogs with Sony, Panasonic and Cannon managers.

Based on the data obtained from field studies, the author first produced an in-depth cases concerning the companies mentioned above. Next, based on this study, he performed analyses and observations from the viewpoint of boundary management. Various scholars (Eisenhardt 1989; Pettgrew 1990; Tin 1994) have discussed the validity of case studies. Case studies make it possible to explain the relevance and cause-and-effect relationships of a variety of observations through deep and detailed insights with consideration given to qualitative information and subjectivity resulting from the peculiarities of individual cases and the difficulties of general analyses. Case studies not only compensate for the weaknesses of generalities but are also indispensable in new, creative theorization.

References

Atkinson, P., Hammersley, M. (1994). Ethnography and participant observation, *Handbook of Qualitative Research*. Denzin, N. K. and Lincoln, Y. S. (eds.),. Thousand Oaks, CA: Sage Publications, 105–117.

Eisenhardt, K. (1989). Building theories from case study research. *Academy of Management Review*, 14, 532–550.

Glaser, B., Strauss, A. (1967). *The Discovery of Grounded Theory: Strategies for Qualitative Research*. Chicago, IL: Aldine.

Lalle, B. (2003). The management science researcher between theory and practice. *Organization Studies*, 24(7), 1097–1114.

Locke, K. (2001). *Grounded Theory in Management Research*. Thousand Oaks, CA: Sage.

Marshall, C., Rossman, G. (1989). *Designing Qualitative Research*. London: Sage.

Pettgrew, A. M. (1990). Longitudinal field research on change: theory and practice. *Organization Science*, 1(1):267–292.

Silvermann, D. (2000). *Doing Qualitative Research*. Thousand Oaks, CA: Sage.

Strauss, A. (1987). *Qualitative Analysis for Social Scientists*. New York: Cambridge University Press.

Yin, R. K. (1994). *Case Study Research: Design and Methods*, 2nd ed. London: Sage.

Index

A

Ackoff, R., 64
ALADIN, 86, 87, 88
Aoki, Dr. M., 46

B

Ba, 9, 94, 95, 157
Baden-Fuller, C., 106
Baldwin, C.Y., 16
Barley, S., 65
Barney, J., 25, 143
Benson, J., 108, 109
Blue Ocean-type strategies, 2
Bohm, D., 125
Boundaries congruence, 7–10, 11, 12, 17,
    30, 33, 49, 51, 53, 54, 55, 58–61, 85–89,
    125–139, 141, 142, 144, 145, 147,
    150, 161
Boundaries consolidation capability, 130,
    131, 133
Boundaries vision, 141, 142
Boundary management, 7–10, 11–12, 15–33,
    37–61, 161, 166
Boundary objects, 45
Brown, J., 95
Burgelman, R.A., 11, 37
Burt, S., 156
Business architecture, 9, 10, 11, 12, 16–23,
    24, 26, 33, 58–61, 85–89, 125, 132,
    133, 141, 147–150, 151, 161–162, 163,
    164
Business ecosystems, 3–5, 16, 32
Business models, 1–5, 7, 8, 12, 15, 16, 17, 18,
    19, 20, 21, 22, 24, 30, 37–61, 63, 64, 66,
    70, 74, 79, 83, 85, 88, 89, 93, 116, 118,
    119, 120, 121, 122, 123, 133, 138, 141,
    142–143, 144, 147, 156, 157, 161, 162,
    163, 164

C

Carlile, P., 10, 57, 125
Chakravarthy, B., 3, 25, 26
Chandler, A.D., 143
Chesbrough, H., 4, 143
Christensen, C.M., 2, 29, 64, 118
Clark, K.B., 16
Coase, R.H., 143
Co-evolution, 3, 4, 50, 143
Coleman, J., 9
Collaborative leadership, 17, 22, 43, 59, 60,
    61, 89, 111, 124
Community leaders, 12, 93, 95, 96, 100, 101,
    105, 106, 107, 108–110, 111–112
Community of practice, 9, 126, 148
Competency traps, 64
Construction innovations, 50–52
Constructive conflict, 109, 150
Content innovations, 54–56
Context architect capability, 130, 131, 133
Context-based organization, 129
Core rigidities, 64
Corporate boundaries, 7, 9, 15–16, 17, 23, 24,
    25, 28, 58, 61, 141, 161
Corporate strategy, 5, 10, 11, 15, 16, 27, 30,
    141, 142, 143, 144, 161
Corporate strategy streams, 10, 11
Corporate system, 7, 8, 9, 12, 16, 17–18, 24,
    25–26, 28, 30, 31, 32, 33, 38, 49, 51, 53,
    54, 58, 59, 60, 63, 65, 85–87, 88, 89, 132,
    133, 141, 142, 144, 145, 147, 148, 150,
    161–162, 163
Creative abrasion, 126
Creative collaboration, 56, 126, 127, 128
Creative destruction, 64
Creativity-based view, 143
Cross-functional teams (CFT), 84, 122, 127,
    130, 133, 145, 146, 154, 155

D
Day, G., 2–3, 127
Dell, 2, 19, 21
Dialectical dialogue, 101, 108, 109, 110
Dialectical leadership, 17, 22, 59, 60, 61, 84,
    85, 86, 88, 89, 95, 96, 106, 107, 108–112,
    124–125, 128
Dialectic-based view, 144
DiMaggio, P., 96, 106
Disciplined imagination, 64, 149
Disruptive technologies, 2
Dougherty, D., 8, 64, 125
DREAMS, 86, 88
Dual network, 12, 63, 65, 74, 78, 81, 84,
    131–139, 150, 154
Dyer, J., 9
Dynamic capability, 25, 29
Dynamic strategic management theory, 5,
    24, 25
Dynamic view of strategic management, 1–12

E
Eisenhardt, K., 10, 15, 115, 142, 147, 166
Embeddedness, 96, 106, 107
Emergent organizations, 147, 148, 149
Engels, F., 109
Environment adaptive strategy, 8, 9, 11, 18,
    25–30, 31, 32, 33, 58–59, 60, 61, 64, 65,
    69, 74, 81, 85, 86–87, 88, 89, 115, 128,
    141, 151
Environment creation strategy, 8, 9, 11, 18, 28,
    30–31, 32, 33, 58–59, 64, 65, 69, 74, 81,
    85–87, 88, 89, 116, 128, 141, 142, 151
Exploitation, 63–65, 74
Exploitative networks, 65, 68, 69, 73–74, 81,
    82, 83, 84
Exploration, 63–65, 73
Exploratory networks, 65, 66, 68, 69, 72–78,
    80–81, 84, 88, 89
External network architecture, 155–158

F
Fast Retailing, 138–139
Flat and web organization, 151
Forceful leadership, 110, 111

G
Ghoshal, S., 43, 144
Giddens, A., 65, 152
Glaser, B., 166
Govindarajan, V., 21
Granovetter, M., 96, 156
Grant, R., 106, 143
Greanleaf, R., 111

H
Hagel, J.III., 126
Hamel, G., 1, 30, 33, 64, 141
Hegel, G.W.F., 109
Henderson, R., 18, 29, 64
Honda, 16, 33, 119, 121, 123, 134, 146, 152,
    166

I
I-mode, 12, 19, 32, 42, 52, 65, 66, 68, 69,
    71, 72, 73, 75–77, 78, 79, 80–81, 82, 86,
    87–88, 89, 156, 157, 158
Innovative leadership, 128, 131
Integrated organization, 12, 17, 20, 21, 47, 145,
    146, 147–150
Integrated video transmission system, 93,
    103–104, 105, 107, 110
Internal corporate venture (ICV), 12, 17, 20,
    37–38, 119, 122, 124, 127
Involvement, 40, 96, 106–107

J
Johansson, F., 64, 127

K
Kanter, M.R., 124, 126
Kaplan, R., 6
KDDI, 4, 29, 78, 81
Kim, W.C., 1, 2, 80, 118
Knowledge boundaries, 8, 9, 10, 11, 37–38, 39,
    116, 126, 127, 128–131, 132, 133, 142, 144
Knowledge creation process, 11, 95, 96, 106,
    165, 166
Kodama, M., 2, 3, 4, 5, 7, 8, 9, 16, 18, 19, 20,
    21, 22, 30, 31, 37, 42, 44–45, 48, 64, 67,
    68, 70, 74, 80, 84, 85, 94, 95, 96, 102, 107,
    116, 118, 120, 124, 125, 126, 128, 131,
    133, 141, 143, 144, 145, 146, 149, 151,
    156, 157
Kogut, B., 15
Kutaragi, K, 39–40, 42, 43, 45, 80

L
Leadership-based strategic community, 12,
    93–112
Leadership teams (LT), 43, 44, 45, 46, 65, 74,
    82, 83–85, 145, 146, 148, 149–150, 154,
    155
Leonard-Barton, D., 43, 64, 126

M
Management drivers, 142, 143, 144
March, J., 63, 64
Marketing innovations, 52–54, 63, 64, 82, 85,
    118

Markides, C., 2, 21, 31, 63
Mental models, 8, 37, 125
Mintzberg, H., 7, 31, 64, 83, 94
Montgomery, C., 5
Moore, J., 4, 32

N
Nadler, D.A., 26, 115
Nelson, R., 15, 63
Network architecture, 12, 142, 144, 151–152, 153–158, 162–164
Network architecture thinking, 142, 144, 151–152, 162–164
Network connectivity, 74, 84, 132, 133, 142, 147, 150
Network effect, 3
Network externality effects, 144
Network integrative competences, 74, 83–85
Nintendo, 2, 38, 39, 40, 41, 42, 46, 118
Nisbett, R., 108, 109
Nonaka, I., 9, 20, 56, 83, 87, 94, 95, 120, 131, 143
Noria, N., 20
NTT-DATA, 12, 20, 21, 32, 33, 37–61, 119, 121, 145, 146, 152
NTT DoCoMo, 4, 12, 19, 20, 29, 32, 33, 42, 52, 63–89, 100, 101, 103, 104, 105, 119, 127, 144, 145, 146, 156, 157, 165, 166

O
Obihiro Project, 100, 105, 106, 107, 109, 110
Omron, 119, 122, 123, 136–137, 152
Open innovation, 4, 143
O'Reilly, C.A., 10, 26
O'Reilly, C. III., 21, 63, 107, 141, 145, 146
Organizational identity, 144
Organization architecture, 145–147

P
Peng, K., 108, 109
Penrose, T., 15, 143, 144
Peripheral vision, 2, 127, 131, 142
Pettigrew, A., 56
PlayStation, 19, 38, 39, 40, 41–42, 43, 44, 45, 46, 58, 59, 77, 80, 118
Poole, M.S., 109
Porter, M., 6, 7, 24, 25, 64, 94, 142
Positioning theory, 24, 25, 30
Positive feedback, 3, 42, 70, 76, 77
Powell, W., 95, 96, 106
Practical knowledge, 38, 56–58, 60, 61, 84, 89, 124
Practice processes, 8, 18, 32, 38, 57, 129, 131
Prahalad, C.K., 1, 4, 30, 64, 141

Process-based organization, 116, 127, 128–132, 133–134, 142, 150–152, 153, 154, 163
Productive friction, 126
Project networks, 46, 70, 72, 73–74, 78, 80, 83, 84, 124, 127, 130, 133

Q
Qualitative research methodology, 165

R
Raynor, M., 2
Real-option strategies, 27
Resonance of values, 96, 106, 107, 110, 124
Resource-based views, 24–25, 27, 94
Robbins, S., 109

S
Sawhney, M., 95, 143
S-curve, 29, 30–31, 80, 88
Sense making, 141, 142
Servant leadership, 111
Shapiro, C., 3, 42
Small-world networks (SWN), 9, 38, 65, 71–72, 74, 81–82, 83, 153–154
Small-world structure (SWS), 9, 11, 12, 17, 18, 21, 22, 38, 43–46, 48, 49, 50, 53, 54, 55, 56–57, 59–60, 65, 74, 93, 100, 108, 123, 124, 125–128, 131, 150–151, 153–158, 163
Softbank, 4, 78
Sony, 4, 12, 15, 19, 20, 21, 27, 33, 37–61, 80, 118, 119, 144, 145, 146, 152, 158, 166
Sony Computer Entertainment (SCE), 19, 21, 22, 38, 39, 40, 41, 42–45, 46, 56, 58, 59, 80
Spears, L., 111
Spender, C., 64, 125
Strategic community-based organizations, 110–112
Strategic community (SC), 12, 93–112, 146
Strategic intent, 30, 64, 141, 142
Strategic management as practice, 5–7
Strategic management as practice process, 5–7
Strategic positions, 1–3, 6, 15, 24, 25, 27, 63, 117
Strategy as practice, 7, 10, 56
Strauss, A., 166
Strogatz, S., 9, 11, 153, 154
Strong ties, 154, 155, 156–157
Structural hole, 156, 157
Structural nature of SWS, 157–158
SWS horizontal integration, 158
SWS relationships, 155–157
SWS vertical integration, 157–158

Synthesizing capability, 95, 96, 106, 107,
    108–110, 111, 112

T
Takara, 119, 122, 137–138, 152, 166
Teece, D., 25, 29
Telemedicine system, 93, 95, 100, 101, 102,
    104, 105, 107
Thought worlds, 8, 37, 64, 125
Time-pacing, 115
Toyota, 16, 33, 52, 119, 135–136, 146, 152,
    166
Traditional organizations, 48, 127, 147, 148,
    149, 150
Transaction cost economics view, 143
TSMC, 19, 20, 21, 22
Tushman, M.L., 10, 26, 115, 116

V
Van de Ven, A.H., 109
Von Hippel, E., 37

W
Walsh, J.P., 144
Watts, J., 9, 11, 153, 154
Weick, K.E., 64, 141
Welch, J., 6
Wenger, E.C., 9
Wernerfelt, B., 25, 143
Whittington, R., 10, 56
Williamson, O.E., 15, 143
Winter, S., 15, 63

Z
Zook, C., 1, 2